C000092213

Stress-Inducible Processes in Higher Eukaryotic Cells

Stress-Inducible Processes in Higher Eukaryotic Cells

Edited by

Thomas M. Koval

National Council on Radiation Protection and Measurements
Bethesda, Maryland

Plenum Press • New York and London

Library of Congress Cataloging-in-Publication Data

Stress-inducible processes in higher eukaryotic cells / edited by
Thomas M. Koval.
    p.   cm.
    Includes bibliographical references and index.
    ISBN 0-306-45677-X
    1. Eukaroytic cells.   2. Stress (Physiology)   I. Koval, Thomas M.
    [DNLM: 1. Stress--physiopathology.   2. Eukaryotic Cells-
    -physiology.   3. Adaptation, Physiological.   QZ 160 S9148 1997]
    QH650.S77 1997
    571.6--dc21
    DNLM/DLC
    for Library of Congress   97-36823
                                                                     CIP

ISBN 0-306-45677-X

© 1997 Plenum Press, New York
A Division of Plenum Publishing Corporation
233 Spring Street, New York, N.Y. 10013

http://www.plenum.com

10 9 8 7 6 5 4 3 2 1

All rights reserved

No part of this book may be reproduced, stored in a retrieval system, or transmitted in
any form or by any means, electronic, mechanical, photocopying, microfilming,
recording, or otherwise, without written permission from the Publisher

Printed in the United States of America

Contributors

EIKO AKABOSHI • Institute for Molecular and Cellular Biology, Osaka University, 1-3 Yamadaoka, Suita 565, Japan

ADNAN ALI • Department of Biology, University of Waterloo, Waterloo, Ontario N2L 3G1, Canada

E. I. AZZAM • Radiation Biology and Health Physics Branch, Atomic Energy of Canada Limited, Chalk River Laboratories, Chalk River, Ontario K0J 1J0, Canada

ROLLIE J. CLEM • Department of Molecular Microbiology and Immunology, The Johns Hopkins School of Hygiene and Public Health, Baltimore, Maryland 21205

S. M. DE TOLEDO • Radiation Biology and Health Physics Branch, Atomic Energy of Canada Limited, Chalk River Laboratories, Chalk River, Ontario K0J 1J0, Canada

R. HAMMERSCHMIDT • Department of Botany and Plant Pathology, Michigan State University, East Lansing, Michigan 48824

JOHN J. HEIKKILA • Department of Biology, University of Waterloo, Waterloo, Ontario N2L 3G1, Canada

YUTAKA INOUE • Osaka University of Foreign Studies, Minoo, Osaka 562, Japan

THOMAS M. KOVAL • National Council on Radiation Protection and Measurements, Bethesda, Maryland 20814

R. E. J. MITCHEL • Radiation Biology and Health Physics Branch, Atomic Energy of Canada Limited, Chalk River Laboratories, Chalk River, Ontario K0J 1J0, Canada

NICK OHAN • Department of Biology, University of Waterloo, Waterloo, Ontario N2L 3G1, Canada

MELVIN J. OLIVER • Plant Stress and Water Conservation Unit, USDA-ARS, Lubbock, Texas 79401

MILTON J. SCHLESINGER • Department of Molecular Microbiology, Washington University School of Medicine, St. Louis, Missouri 63110

YING TAM • Department of Biology, University of Waterloo, Waterloo, Ontario N2L 3G1, Canada

GAYLE E. WOLOSCHAK • Center for Mechanistic Biology and Biotechnology, Argonne National Laboratory, Argonne, Illinois 60439

ANDREW J. WOOD • Plant Stress and Water Conservation Unit, USDA-ARS, Lubbock, Texas 79401; *present address*: Department of Plant Biology, Southern Illinois University at Carbondale, Carbondale, Illinois 62901

Preface

Stress in various forms confronts all living organisms. Evolution is indebted in large part to the selective pressures created by stresses. Types of stress range from omnipresent agents, such as the background radiation that is part of the earth's natural environment, to quite specific agents or conditions that affect only certain organisms or cell types. Even normal physiological processes can be stressful as evidenced by the large amount of DNA damage induced via cell metabolism. Therefore, it is not a surprise that eukaryotic as well as prokaryotic cells have evolved numerous methods of countering both intrinsic and environmental stresses. The mechanisms utilized by eukaryotic cells in responding to stresses are remarkably similar from lower plants through humans as will be evident after reading the chapters of this book.

Until about the last quarter of this century, it was common for scientists to be true intellectuals who were conversant in a variety of academic areas and also possessed global knowledge within their specific field of interest. With the knowledge explosion of recent decades, it has become more difficult for scientists to be versed in multiple academic areas and, in fact, to be knowledgeable about the broad aspects of their own research area. Current scientific journal titles reflect the degree of specialization of contemporary scientists. This specialization has its rewards in fantastic accumulations of information in very focused areas of research. However, this specialization has as its expense the isolation of scientists into very limited niches. It is hoped that this volume can serve to expand the perspectives of those scientists involved in the various specific aspects of the stress biology of eukaryotic cells as well as provide a source of potentially unifying concepts and hypotheses for students interested in the evolutionary basis of biology.

The goal of this book is to provide an overview of cellular stress responses with an emphasis on diversity in both the types of stress and the biological systems studied. The types of stress range from ionizing and ultraviolet radiations to alkylating agents, heat, virus, and desiccation. Systems examined represent plants, invertebrates, lower vertebrates, and mammals, including humans. The broad scope of work delineated in the various chapters allows the reader to examine the similarities among inducible responses initiated by a variety of agents in this wide cross section of eukaryotic systems, ranging from plants to humans. The gathering of such specialized diverse studies in this single volume will hopefully serve as an interdisciplinary forum in bringing together areas of research that may not otherwise have found common ground.

This book contains contributions from nine different authors or groups. This gives the reader the distinct benefit of obtaining the most recent information available in each specific area as presented by those intimately involved therein. For such a diverse group of topics as is included in this volume, this is a special advantage, one that would be extremely difficult for a single author to provide. It is hoped that the somewhat heterogeneous writing styles and minor repetitions and inconsistencies can be overlooked for this reason.

The chapters of this book are organized in an evolutionary progression from lower to higher organisms. (In these days of political correctness, I must apologize to any moss or plants or their champions who are offended at my presumption that these are the lowest organisms treated in this volume.) Chapters 1 and 2 discuss stress responses in plants. In Chapter 1, Drs. Oliver and Wood discuss a somewhat unusual stress confronting some organisms, namely, desiccation. Their illustration of how certain bryophytes (mosses) are able to cope with this extremely severe condition provides an interesting beginning to this book. Chapter 2 follows with a description of acquired stress resistance in plants. Dr. Hammerschmidt presents a summary of infection processes, defensive responses, and endogenous signals for resistance in plants. Chapter 3 switches to animals with a discussion of stresses in *Drosophila*, one of classical biology's most popular systems. Drs. Akaboshi and Inoue provide information on the responses induced by a number of different agents in the cells of this dipteran insect, which has played a leading role in delineating the heat stress response. Chapter 4 continues with another insect order, the Lepidoptera. Dr. Koval examines the pronounced resistance of these cells to many physical and chemical agents and the associated induced responses. Chapter 5 concentrates on cell death as an active stress response. Dr. Clem describes baculovirus studies in lepidopteran cells, providing convincing evidence of the importance of apoptosis in the survival of organisms and its conservation in metazoans ranging at least from insects and nematodes to humans. Chapter 6 proceeds to amphibians where Drs. Heikkila, Ali, Ohan, and Tam relate the importance of stress proteins in normal cells as well as those subjected to environmental and chemical agents. They discuss the molecular chaperone and protein sequestering roles of the stress proteins. Chapter 7 elevates us to avian systems and focuses primarily on heat stress. Dr. Schlesinger has been particularly sensitive to the potential complications of necrosis in his discussion of avian cells, and details the importance in experiments of adjusting stress conditions to allow for virtually complete recovery of cells after the stress has been removed. The final two chapters concentrate on mammalian cells. In Chapter 8, Dr. Woloschak presents an overview of radiation-induced responses in mammalian cells, factors influencing the responses, and mechanisms of response induction. In Chapter 9, Drs. Mitchel, Azzam, and de Toledo describe the adaptive response and demonstrate this phenomenon in both rodent and human cells in response to ionizing radiation exposures.

I want to express my gratitude to the contributors of this volume for accepting my invitation to participate in this effort and for producing such excellent chapters. I also appreciate their cooperation in making suggested revisions and tolerating my editorial alterations. I hope this book will be helpful to researchers, teachers, and students interested in the evolutionary aspects of the stress response in eukaryotic cells.

Thomas M. Koval

Contents

xi

3. Stress Responses in *Drosophila* Cells

EIKO AKABOSHI and YUTAKA INOUE

4. Stress Resistance in Lepidopteran Insect Cells

THOMAS M. KOVAL

Desiccation Tolerance in Mosses

MELVIN J. OLIVER and ANDREW J. WOOD

1. INTRODUCTION

To gain a full understanding of stress-inducible processes in plants, especially at the cellular level, it is often of major benefit to develop simple model plants for study. This is especially true if one is interested in how plants tolerate extremely stressful conditions that impact directly on the protoplasm of individual cells, e.g., desiccation. In addition, many crop species have little capacity for abiotic stress tolerance and thus the genetic information necessary for expanding their tolerance may not be present or exploitable. Model plants that exhibit stress-tolerant traits are useful tools for elucidating the processes involved in tolerance and may provide unique genetic material that can impact breeding programs for improved crop stress management.

Cell cultures derived from vegetative or reproductive plant

This chapter was prepared by a U.S. government employee as part of his official duties and legally cannot be copyrighted.

MELVIN J. OLIVER and ANDREW J. WOOD • Plant Stress and Water Conservation Unit, USDA-ARS, Lubbock, Texas 79401. *Present address of A.J.W.*: Department of Plant Biology, Southern Illinois University at Carbondale, Carbondale, Illinois 62901.

Stress-Inducible Processes in Higher Eukaryotic Cells, edited by Koval. Plenum Press, New York, 1997.

tissues have proven useful in the study of some abiotic stress responses but cells in culture often do not respond as cells maintained within a tissue. Alternatives that maintain simplicity and furnish cells in a "native" structure are better suited and offer a more realistic model for stress-induced cellular responses in more complex crop species. Bryophytes are such model plants, in particular the vegetative gametophytic generation of many mosses. Plants of this group are exceedingly simple but have enough similarity to more complex plant tissues to be useful. They are composed of leafy stems that in many mosses consist of an axial conducting core (nonvascular), small rhizoids, and leaves that in most species are one cell thick (except along a well-defined nonvascular midrib). The one-cell thickness of the leaves allows for the almost instantaneous application of a stress to the protoplasm, and as most are composed of only one major cell type the effect approaches homogeneity. Such attributes have proven extremely useful in the study of atmospheric pollutants and, at least in the case of desiccation and oxidative agents, are helping to unravel the intricacies of stress-tolerance mechanisms at the cellular level in plants. As an added bonus, moss gametophytes are haploid, which, as transformation procedures become more common, offers a great advantage for genetic and molecular-based investigations into stress-induced responses. This chapter is centered on the role of mosses as model plants for the study of desiccation tolerance and how such studies have impacted our knowledge of the cellular basis of stress-tolerance mechanisms in plants, in particular desiccation.

2. DESICCATION TOLERANCE

Desiccation tolerance is the ability of cells to revive from the air-dried state (Bewley, 1979). A wide variety of organisms are capable of surviving desiccation ranging from bacteria to the much more complex multicellular insects and arthropods. Most plants are capable of producing desiccation-tolerant tissues, in particular seeds and pollen, but few can survive the drying of their vegetative structures. Even though such plants are few in number, individual species represent most major classes (Fig. 1). Representatives of the more complex groups of plants, 60–70 species of ferns and fern allies and at least 60 species of angiosperms (Bew-

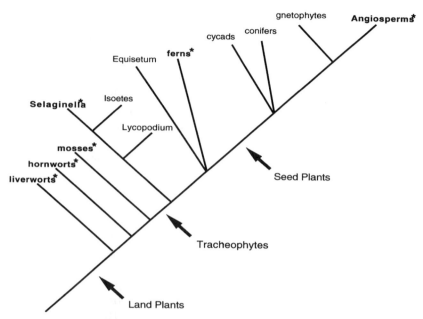

FIGURE 1. An outline phylogeny of the land plants, based on a consensus of several recent synthetic studies (Crane, 1990; Donoghue, 1994; Mishler *et al.*, 1994). Names in bold and with an asterisk indicate clades with some known desiccation-tolerant members (data taken from Bewley and Krochko, 1982). Parsimony would suggest at least one independent evolution (or reevolution) of desiccation tolerance in *Selaginella*, the ferns, and the Angiosperms.

ley and Krochko, 1982), require that water loss be a slow process in order to induce and establish mechanisms for tolerance of desiccation (see discussion below), and have been classified as modified desiccation-tolerant plants (Oliver and Bewley, 1997). Such plants have morphological and physiological mechanisms that retard the rate of water loss to the extent required to establish tolerance. Many of the desiccation tolerant plants that are found in the phylogenetically basal clades that constitute the algae, bryophytes, and lichens are truly desiccation tolerant as their internal water content rapidly equilibrates to the water potential of the environment. These plants have very little in the way of water retaining morphological or physiological characteristics and thus the rate of water loss depends on the water status of the immediate environment. As a result of this, many lichens, algae, and desert bryophytes experience drying rates that are extreme, i.e., where their tissues reach air dryness within an hour. Such rates would prove lethal to desiccation-tolerant ferns or angiosperms.

What are the proposed cellular mechanisms by which plant cells achieve desiccation tolerance? Bewley (1979) established three criteria that a plant or plant structure must meet to survive desiccation. It must: (1) limit the damage incurred to a repairable level, (2) maintain its physiological integrity in the dried state (perhaps for extended periods of time), and (3) mobilize repair mechanisms on rehydration that effect restitution of damage suffered during desiccation (and on the inrush of water back into the cells). These criteria can be simplified into two basic components by which desiccation tolerance can be achieved: the protection of cellular integrity and the repair of cellular damage induced by desiccation (or rehydration), as described by Bewley and Oliver (1992). Plants, in all probability, employ mechanisms that encompass both components, but as desiccation tolerance has evolved independently on a minimum of 12 separate occasions (Oliver and Bewley, 1997), one would expect that there exist examples of desiccation-tolerant plants that span the spectrum of possible combinations of the two strategies; from plants or structures that rely heavily on cellular protection to those that rely more on cellular repair. Work with desiccation-tolerant seeds and modified desiccation-tolerant angiosperms leads one to the conclusion that these tissues and plants utilize mechanisms that rely heavily on inducible (either developmentally programmed for seeds or environmentally induced in vegetative tissues) cellular protection systems (for reviews see Bartels and Nelson 1994; Bartels et al., 1993; Bewley et al., 1993; Bewley and Oliver, 1992; Burke, 1986; Close et al., 1993; Crowe et al., 1992; Dure, 1993; Gaff, 1989; Leopold et al., 1992; Oliver and Bewley, 1997). The mechanisms of tolerance that are used by seeds and modified desiccation-tolerant plants appear to involve two major components, sugars and proteins, both of which are postulated to be involved in protection of cellular integrity during the drying phases (Bewley et al., 1993; Crowe et al., 1992; Dure, 1993; Leopold et al., 1992; Oliver and Bewley, 1997). The induction and establishment of such protective components is thought to be why modified desiccation-tolerant plants do not survive rapid water loss. It thus appears that the speed of desiccation may have important consequences for the type of mechanism by which plants achieve vegetative desiccation tolerance.

Plants that experience rapid desiccation rates have to rely on mechanisms that are either constitutive to the hydrated state, quickly induced during the first phases of water loss, or induced when the stress is relieved by the readdition of water (or perhaps a combination of all three). Such mechanisms may be reduced or not present in plants that cannot withstand rapid drying and thus novel insights into plant stress responses and perhaps basic cellular repair mechanisms may be gained by the study of highly tolerant species. Indeed, as this chapter will demonstrate, desiccation-tolerant plants that withstand rapid rates of drying, in particular the moss *Tortula ruralis*, indicate that such plants employ a level of constitutive protection in conjunction with a rehydration-induced recovery system to establish tolerance (Bewley and Oliver, 1992; Bewley *et al.*, 1993; Oliver, 1996; Oliver and Bewley, 1997). These studies that are the main focus of this chapter and our discussion of the effects of desiccation on plant cells will be limited to studies involving mosses as model plants and how such investigations have broadened our understanding of stress responses in plants.

2.1. Alterations in Ultrastructure

Although there is not a wealth of information concerning the effects of water loss on cellular integrity, it appears that desiccation-tolerant plants do not exhibit any major differences in the type of damage incurred from that seen in desiccation-intolerant tissues or in nontolerant tissues exposed to sublethal water stress (Bewley and Krochko, 1982; Oliver and Bewley, 1984a). Nevertheless, it is important to ascertain when in a wet/dry/wet cycle cellular damage takes place so as to gain an insight into the overall strategy employed as a mechanism for desiccation tolerance, i.e., where in the spectrum from protection to repair the mechanism occurs.

The majority of studies into the structure of dried vegetative tissues have been conducted using gametophytic tissues of desiccation-tolerant bryophytes, in particular mosses (see Oliver and Bewley, 1984a, for review). Nonultrastructural observations using Nomarski optics, for which fixation is not required, demonstrate that *Tortula* species undergo extensive plasmolysis on desiccation. In *T. ruralis*, protoplasm condenses at the proximal and distal ends of the cell leaving the central portions of the cell empty

(Tucker *et al.*, 1975). The regions of condensed protoplasm are connected by cytoplasmic bridges that extend along the abaxial, adaxial, and lateral sides of the dehydrated cells. The chloroplasts in these cells are smaller and more rounded than those in hydrated cells and the nucleus appears unaffected by drying. Not all desiccation-tolerant mosses exhibit this type of plasmolysis. Drying cells of *Triquetrella papillata* and *Barbula torquata* contract to 50 to 70% of their original volume at desiccation, which apparently excludes the entry of air through the cell walls. This results in the protoplasm of these cells occupying the total area enclosed by the cell walls (Moore *et al.*, 1982). These studies demonstrate that drying elicits major changes in cell structure but do not directly demonstrate cellular damage. Such information is best gained from more detailed ultrastructural studies. The results from early studies of fixed dried cells of desiccation-tolerant mosses suggest that desiccation elicits shrinkage of organelles, coupled with a breakdown of internal membrane structures such as thylakoids and cristae, and of cytoplasmic membranes such as the endoplasmic reticulum (reviewed by Oliver and Bewley, 1984a). However, there is a caveat to the conclusions drawn from electron micrographs of fixed dried plant cells. Fixation procedures, for however brief a time, always result in some rehydration, which in all probability results in damage to the dried tissues. Thus, it is difficult to assess the effect drying per se has on cellular integrity. Recently, Platt *et al.* (1994) attacked this problem using freeze-fracture technology, which avoids rehydration, to establish the condition of plasma membranes and organellar membranes in dried leaf cells of *T. ruralis* and *Selaginella lepidophylla* (a modified desiccation-tolerant fern). The cell membranes of both species in the dried state remain as intact bilayers containing normally dispersed intramembranous particles. The structural organization of the organelles is also maintained in both species, both thylakoid and crista membranes appear intact, and no areas of disrupted bilayer organization were detected in any of the cell membranes. Thus, it appears that for *Tortula* and *Selaginella* membrane disruption does not occur during drying but during subsequent rehydration. Similar conclusions have been drawn from the use of freeze-fracture technology to investigate membrane structure in dried seeds (Bliss *et al.*, 1984; Platt-Aloia *et al.*, 1986; Thomson and Platt-Aloia, 1982). It is therefore possible that this may be a general rule for dry plant tissues, and if so, it could be inferred that protection mechanisms, whether induced or constitutive, are very successful in maintaining membrane structures during the removal of water from plant cells.

Regardless of the ability to protect membranes during drying, all desiccation-tolerant vegetative plants and tissues exhibit cellular damage immediately on rehydration, either as a result of the inrush of water or as a consequence of the drying process, or both. That the drying process does influence cellular damage is suggested by the observation that rapidly (within an hour) dried *T. ruralis* phyllidial cells suffer a greater degree of disruption on rehydration than do the same cells dried slowly (Oliver and Bewley, 1984b; Schonbeck and Bewley, 1981). Thus, although drying does not result in visible injury to membranes, there is some component of the drying process that affects the capability of the membranes to withstand the rigors of rehydration.

The most evident symptom of cellular damage in all desiccation-tolerant plant tissues is the leakage of cellular ions into the surrounding water during and immediately following rehydration. Such leakage has been used as a measure of the extent of one aspect of cellular disruption, i.e., membrane damage, resulting from the desiccation/rehydration event (for reviews see Bewley, 1979; Bewley and Krochko, 1982; Gaff, 1980; Simon, 1978; Simons and Mills, 1983). In desiccation-tolerant bryophytes the majority of the leakage of ions occurs in the first 5 min of reexposure to water and the amount of ions leaked is dependent on the speed at which drying occurred prior to rehydration (Bewley and Krochko, 1982; Oliver and Bewley, 1984a; Oliver *et al.*, 1993) Again this indicates that the drying phase has an impact on cellular damage manifested on rehydration, perhaps time is required even in these desiccation-tolerant plants for full protective measures to be implemented. In the minutes following the early outrush of ions, desiccation-tolerant bryophytes reabsorb the cellular nutrients lost to the surrounding water, which is indicative of a rapid return of the phyllodial cell membranes to their normal selectively permeable state. Such a recovery does not occur in rehydrating cells of desiccation-intolerant species (Oliver and Bewley, 1984a).

As water enters the dried cells of *T. ruralis*, the condensed cytoplasm rapidly expands to fill the empty cell cavity formed by plasmolysis (Tucker *et al.*, 1975) Within minutes after rehydration, chloroplasts are swollen and globular in shape and their outer membranes are folded and separated from the thylakoids, which themselves are no longer compacted (Bewley and Pacey, 1978; Tucker *et al.*, 1975). The extent of thylakoid disruption is dependent on the prior speed of desiccation; the more rapid the drying rate the more disruption occurs. Mitochondria also swell and exhibit disruption of the internal membrane structures (cristae), but the appearance of this organelle on rehydration is not affected by the rate of desiccation. Similar results have been reported for other desiccation-tolerant moss species (Moore *et al.*, 1982; Noailles, 1978; Swanson *et al.*, 1976). In all cases organelles regain normal structure within 24 hr of the readdition of water. Rehydrated cells of dried gametophytes of the desiccation-sensitive moss *Cratoneuron filicinum* exhibit identical structural abnormalities as those seen in *T. ruralis* but in this case the cells never regain a normal appearance and die (Bewley and Pacey, 1978; Krochko *et al.*, 1978). These studies attest to the fact that cells of desiccation-tolerant bryophytes are not immune to the rigors of drying and rehydration. It is their ability to recover from such damage that makes them unique and useful as model plants for the study of stress-induced cellular processes.

2.2. Alterations in Metabolism

2.2.1. Photosynthesis and Respiration

Early research on bryophyte physiology (in the 1970s and early 1980s) focused on the fundamental effects of desiccation and rehydration on photosynthesis and respiration. This body of work has been extensively reviewed (Bewley, 1979; Bewley and Krochko, 1982) and hence will only be briefly discussed here. Later work has been concerned with the ecophysiological and microdistributional aspects of desiccation-induced changes in carbon balance and does not directly impact on the thesis of this discussion.

Gaseous exchange in bryophytes occurs freely at the cell surface when the gametophytes are hydrated (stomates are absent in these plants). During drying the ability to photosynthesize and respire diminish and neither occur in the dried state. On rehydration gaseous exchange rapidly resumes in desiccation-tolerant species, but not in desiccation-intolerant mosses (for references see Bewley and Krochko, 1982). The speed at which desiccation occurs has a significant effect on the response of both photosynthesis and respiration. On rehydration of rapidly dried (within an hour) *T. ruralis*, oxygen consumption is elevated approximately twice the level of that seen in controls (Krochko *et al.*, 1979). This phenomenon is termed *resaturation respiration* and is common for desiccation-tolerant bryophytes. Control rates of oxygen consumption in rehydrated rapidly dried *T. ruralis* are not achieved for about 24 hr. Rehydration of slowly dried moss results in only a moderate increase in oxygen consumption and control levels are reachieved sooner, 10–12 hr. Not all of the oxygen consumption during "resaturation respiration" can be attributed to normal respiratory pathways, however, as carbon dioxide evolution, in the absence of photosynthesis, during rehydration is elevated by no more than 10% (Bewley *et al.*, 1978; Krochko *et al.*, 1979). Nonautotrophic carbon fixation can only explain a very small portion of this discrepancy (Sen Gupta, 1977). That the link between oxygen consumption and carbon dioxide evolution is not normal during "resaturation respiration" is not surprising as the complex inner structure of mitochondria is severely disrupted for a considerable time following rehydration (see above and Oliver and Bewley, 1984a). The amount of disruption is less if the moss is dried slowly, which correlates well with the lower "resaturation" respiratory rate and quicker recovery of control levels seen in these plants. ATP levels during rehydration of dried *T. ruralis* gametophytes are relatively stable and do not appear to change with the speed of prior desiccation even though oxygen consumption is considerably higher in previously rapidly desiccated material (Bewley *et al.*, 1978; Krochko *et al.*, 1979). This suggests that ATP synthesis becomes largely uncoupled from electron transport at this time although enough is made to maintain a constant pool. The mechanism by which desiccation and rehydration alter the fundamental pathways of respiration to cause the observed effects is not understood and requires more detailed analyses.

The speed of photosynthetic recovery after rehydration is also dependent on the prior rate of desiccation in *T. ruralis* (Bewley and Krochko, 1982) and other desiccation-tolerant mosses (Alpert and Oechel, 1987). In general, the faster the speed of desiccation, the slower is the recovery of photosynthesis to normal hydrated levels. The time that the moss spends in the dried state is also an important factor in the recovery of photosynthesis on rehydration. Rapidly dried *T. ruralis* kept dry for 2 days exhibited a temporary depression in net oxygen evolution on rehydration, and this depression was significantly greater for moss kept dry for 7 days prior to rehydration (Schonbeck and Bewley, 1981). Such an effect is also seen in rehydrating slowly dried gametophytes but to a lesser extent. The effect of previous desiccation history on the recovery of photosynthesis on rehydration has important consequences for the maintenance of long-term carbon balances in mosses. This in turn appears to dictate the microdistributions of desiccation-tolerant mosses (Alpert and Oechel, 1987).

2.2.2. Oxidative Metabolism

This area of metabolism plays an important role in the overall mechanism of desiccation tolerance. The direct generation of active oxygen species during desiccation and indirectly through the inhibition of photosynthesis during drying is considered to be a major factor in the cellular damage related to this stress in plants. Mechanisms to limit the production of such species and moderate their effects have to play a role in desiccation tolerance and the evidence obtained from studies involving desiccation-tolerant bryophytes strengthens this conclusion. We will briefly discuss these studies here; for a comprehensive review of this area of metabolism and to place these studies in the larger context of stress metabolism in plants, the reader is directed to recent reviews by McKersie (1991) and Smirnoff (1993).

Oxidative damage during desiccation results in several types of cellular damage including the oxidation of protein sulfhydryl groups leading to denaturation, pigment loss and photosystem damage (especially in the light), and lipid peroxidation and free fatty acid accumulation in membranes (McKersie, 1991; Smirnoff, 1993). Protective mechanisms that have been identified in plants involve enzymes such as peroxidase, catalase, and superoxide dismutase (SOD) along with antioxidant compounds such as ascorbic acid, α-tocopherol, carotenoids, and glutathione (GSH) and the processes that regenerate them. All appear to be directed toward reducing the amount of free radicals generated during desiccation.

Oxidative inactivation of sulfhydryl-containing enzymes as a result of desiccation has been reported in several desiccation-tolerant mosses (Stewart and Lee, 1972). This damage continues during storage of the tissues in the dried state and can be alleviated by incubation with GSH (Stewart and Lee, 1972). GSH is present throughout the cell, and in highest amounts in the chloroplast (Smith *et al.*, 1990); it plays an important role in maintaining the appropriate redox level. Slow desiccation of *T. ruralis* results in the oxidation of the GSH pool to approximately 30% of the oxidized form of GSH, GSSG (Dhindsa, 1987), indicating a decreased ability of the moss to withstand oxidative injury. Interestingly, desiccation itself does not result in an increase in GSH oxidation as GSSG does not increase when the moss is dried rapidly (Dhindsa, 1987), but it does increase on rehydration of rapidly dried moss. Desiccation of the tolerant moss *T. ruraliformis* does not result in a loss of GSH (Seel *et al.*, 1992a). How these conflicting findings relate to an overall mechanism of desiccation tolerance is enigmatic but may simply reflect varying capabilities among desiccation-tolerant species to buffer oxidative damage in this way. In *T. ruraliformis* (Seel *et al.*, 1992a) ascorbate decreases during drying; hence, maintenance of high amounts of GSH may be more important in protection in these plants than it is in *T. ruralis*. *T. ruraliformis* also maintains an appreciable α-tocopherol content during drying, but this is depleted in a desiccation-intolerant species, *Dicranella palustris* (Dicks.) Crundw., again indicating that other antioxidants can be more important than ascorbic acid or GSH (Seel *et al.*, 1992a).

In all plant tissues studied to date, light increases the amount of desiccation-induced damage as a result of oxidation (Smirnoff, 1993). Oxidative damage incurred during drying of *T. ruraliformis* and *D. palustris* increases if the plants are irradiated under high light conditions (1100–1200 μmole m^{-2} sec^{-1}). Neither photosynthetic pigment content nor the ability of *T. ruraliformis* to recover was affected, but there was a reduction in pigments in the sensitive species, *D. palustris* (Seel *et al.*, 1992a). Loss of pigments during desiccation is not a feature of desiccation-tolerant bryophytes but it is common in many modified desiccation-tolerant angiosperms, the so-called poikilohydrous, poikilochlorophyllous plants (Bewley and Krochko, 1982).

Lipid peroxidation also occurs during desiccation of vegetative tissues in bryophytes. By measurement of malondialdehyde, an indicator of lipid peroxidation, Seel *et al.* (1992b) demonstrated that the relatively desiccation-intolerant moss *D. palustris* exhibited increased lipid peroxidation following desiccation, whereby the tolerant species *T. ruraliformis* did not. The extent of lipid peroxidation as a whole, whether in hydrated or desiccated gametophytes, was five- to sixfold higher in the intolerant species.

This may be indicative of an inherent protection against lipid peroxidation in tolerant species. Stewart and Bewley (1982) recorded a decrease in lipoxygenase activity (a lipid peroxidation enzyme) during desiccation, again indicating a protective mechanism inherent in desiccation-tolerant species.

2.2.3. RNA and Protein Synthesis

As gametophytic tissues of mosses dry, whether or not they are desiccation-tolerant, the cells' ability to conduct protein synthesis rapidly declines (Bewley, 1972, 1973b; Henckel *et al.*, 1977; Siebert *et al.*, 1976; M. J. Oliver, unpublished data for *T. caninervis* and *T. norvegica*). In *T. ruralis* this loss of protein synthetic capacity is manifested in a loss of polysomes resulting from the run-off of ribosomes from mRNAs coupled with a failure in the initiation machinery (for reviews see Bewley, 1979; Bewley and Oliver, 1992). Rapid desiccation of *T. ruralis*, however, leads to the retention of 50% of the polysomes in the dried state, indicating that water loss alone is not the cause of the detachment of ribosomes from mRNAs. In this case, it is thought that water loss is so fast that mRNAs are trapped on polysomes before run-off is completed. The rapid loss of polysomes during drying (under "natural" drying rates) and the apparent sensitivity of the initiation step of protein synthesis to protoplasmic drying lead us to the conclusion that the induction of synthesis of "protective" proteins during drying is highly unlikely. This is borne out by the observation that no new mRNAs are recruited into the protein synthetic complex even during slow drying (Oliver, 1991). The fact that the moss survives rapid desiccation indicates that an inducible protection mechanism is not necessary for survival. In addition, if protective proteins (or other compounds) are an important part of the desiccation tolerance of this species, they must be present at all times, i.e., constitutively expressed.

Protein and RNA synthesis recover rapidly on rehydration of desiccation-tolerant mosses (Bewley, 1973a,b; Gwozdz *et al.*, 1974; Henckel *et al.*, 1977; Oliver and Bewley, 1984b; Oliver *et al.*, 1993; Siebert *et al.*, 1976). In both cases the rate of recovery to control levels is faster the slower the rate of prior desiccation (Gwozdz *et al.*, 1974; Oliver and Bewley, 1984b). The slower rate of recovery for protein synthesis on rehydration of rapid-dried moss occurs even though polysomes are retained in the dried state. This is thought to be indicative of the greater degree of cellular damage in these gametophytes. In *T. ruralis*, ribosomes and ribosomal RNAs are stable during desiccation and both the conserved pool of these components and those that are newly synthesized quickly embark on the formation of new polysomes on rehydration (Gwozdz and Bewley, 1975; Oliver and Bewley, 1984b,c; Tucker and Bewley, 1976). mRNAs are also stable to desiccation, more so in slow-dried moss than in rapid-dried moss, and are rapidly utilized in protein synthesis on rehydration (Oliver and Bewley, 1984c,d). On rehydration of the dried gametophytes there is a turnover of mRNAs stored in the dried state accompanied by a replenishment of the message pool as a result of *de novo* synthesis (Oliver and Bewley, 1984d). The rate of mRNA synthesis is faster if there is greater cellular disruption as a result of rapid desiccation. This may be in response to the greater loss of mRNA during rapid drying or it could reflect the activation of a repair mechanism to overcome the greater degree of disruption in these cells. The turnover rate of conserved mRNAs on rehydration is unaffected by the desiccation event. However, the proportion of conserved mRNA in polysomal fractions of rehydrated gametophytes, whether dried slowly or rapidly, soon declines such that within 2 hr of rehydration little is present (Oliver and Bewley, 1984d): Most of the mRNA within the activated protein synthetic complex at this time is newly synthesized.

3. GENE EXPRESSION

Desiccation-tolerant plants have only recently received some attention with regard to alterations in gene expression that may be involved in the underlying mechanisms of tolerance. To date five species have been investigated in this manner (for reviews see Oliver and Bewley, 1992, 1997]: the modified desiccation-tolerant angiosperm *Craterostigma plantagineum* (Bartels *et al.*, 1993; Bartels and Nelson, 1994), the modified desiccation-tolerant grass *Sporobolus stapfianus* (Gaff *et al.*, 1992), the modified desiccation-tolerant fern *Polypodium virginianum* (Reynolds and Bewley, 1993), the modified desiccation-tolerant moss *Funaria hygrometrica* (Bopp and Werner, 1993), and the desiccation-tolerant moss *Tortula ruralis* (Bewley and Oliver, 1992; Oliver and Bewley, 1997). The work involving modified desiccation-tolerant plants has been extensively reviewed (Bartels *et al.*, 1993; Bartels and Nelson, 1994;

Bewley *et al.*, 1993; Oliver and Bewley, 1992, 1997) and thus our discussion in this section will concentrate on the extensive analysis of gene expression in *T. ruralis* and the contribution it has made to our understanding of desiccation tolerance and to the response of plants in general to stressful events.

3.1. Alterations in Protein Synthesis (Hydrins and Rehydrins)

As discussed earlier, the rapid loss of polysomes during desiccation coupled with a desiccation-induced failure in initiation of message recruitment argues against the synthesis of novel proteins during drying (indeed it is unlikely that any protein synthesis continues for any length of time after water begins to leave the cells). Direct evidence for this comes from the lack of the synthesis of novel proteins from polysomal complexes isolated from drying gametophytes (Oliver, 1991). In response to rehydration, however, protein synthesis rapidly recovers and the pattern of proteins synthesized is extensively different from that seen under non-stressed conditions (Oliver, 1991; Oliver and Bewley, 1984d). This change in the pattern of protein synthesis occurs without a qualitative change in the pool of mRNA available for translation (Oliver and Bewley, 1984d; Scott and Oliver, 1994), suggesting that the transcriptional responses effecting the synthesis of stress-related proteins that are common in stressed plants (Sachs and Ho, 1986) are not operating in the response of the moss to desiccation. The inference from this study is that the alteration in gene expression associated with rehydration is mediated mainly by an alteration in translational controls, at the level of mRNA selection, and that if transcription is important it serves to replenish the pools of pre-existing messages. Such a scenario is attractive as the moss has to respond quickly to rapid changes in the water status of its environment; the speed of the response would be greater if effected at the level of translation.

In a detailed study of the change in gene expression on rehydration, Oliver (1991) demonstrated that during the first 2 hr the rates of synthesis of almost 80% of the proteins are either increased or decreased significantly from their rates of synthesis observed under continuously hydrated conditions. Of particular note are the proteins whose synthesis is dramatically affected. These include 25 proteins whose synthesis is terminated, or substantially decreased (greater than fivefold), and 74 proteins whose synthesis is initiated, or substantially increased (greater than fivefold). These proteins have been designated *hydrins* and *rehydrins*, respectively. These are functional terms and do not refer to any common sequence motif or structural property nor imply a common enzymatic function. The altered synthesis of these two groups of proteins is not coordinately regulated (Oliver, 1991). Rehydrin synthesis is initiated or stimulated on rehydration of gametophytes that had been previously dried to between 50 and 20% of their fresh weight, while the synthesis of hydrins is steadily inhibited as gametophytes are subjected to greater degrees of water loss prior to rehydration. Their synthesis is almost completely inhibited in rehydrated tissues that were previously dried to 50% of their fresh weight. These findings indicate that the synthesis of rehydrins is fully activated only after a threshold of prior water loss is reached. If this is so, then, at least for *T. ruralis*, the activation of a repair or recovery mechanism requires that a certain level of stress be experienced and to accommodate this there may be a stress level "sensing" component to the response.

Hydrins and rehydrins also differ in the time required to return to normal levels of synthesis during extended periods of hydration. The synthesis of all hydrins returns to control levels between 2 and 4 hr following rehydration. The reduction in synthesis of rehydrins to control levels depends on each individual rehydrin, however. Some rehydrins are only synthesized for less than 2 hr, whereas others are still being synthesized at elevated levels 10 to 12 hr after rehydration. A full return to control levels of synthesis for all proteins is evident after 24 hr. From this study it is possible to classify rehydrins as either early recovery proteins or rehydrins that are required for extended periods during recovery of the moss from desiccation. This observed diversity of rehydrin gene expression is consistent with the idea that *T. ruralis* utilizes a cellular repair-based strategy of tolerance. One would expect that once the initial potentially lethal damage is repaired in the first hours of rehydration (see earlier section on leakage and ultrastructural damage) the longer-term repair and recovery of organelles and associated metabolism would take place. If this notion is correct, then one would predict the early rehydrins to be involved in such processes as the repair of the damage to the plasma membrane (to stop the loss of the cytoplasmic contents) and those whose synthesis is protracted to be involved in the longer-term recovery of organellar function and structure. At this time this is simply a working hypothesis and its proof or disproof will await the identification and characterization of individual rehydrins.

3.2. Alterations in Transcript Availability and Accumulation

To more closely follow rehydrin mRNA pools and gain an insight into the translational controls affected during desiccation and rehydration, Scott and Oliver (1994) isolated 18 cDNAs that correspond to transcripts preferentially translated during rehydration of dried *T. ruralis* gametophytes. Consistent with the earlier findings that the mRNA pool does not change qualitatively during desiccation and rehydration, rehydrin transcripts are not unique to rehydration. All 18 rehydrin cDNAs represent mRNAs present in the gametophytes prior to desiccation but they are present in greater amounts in the polysomes of rehydrated gametophytes compared with those from undesiccated moss. This latter finding supports the hypothesis that it is an alteration in translational controls that drives the induced change in gene expression. Again, it is also suggestive that the alteration in control is at the level of differential selection and/or recruitment by the translational machinery from a qualitatively constant mRNA pool (Oliver, 1991).

In conjunction with the recruitment on rehydration we have also demonstrated that individual rehydrin transcripts accumulate during drying if the rate of water loss is similar to natural drying rates, i.e., slow drying (Fig. 2A,B). This accumulation occurs during and subsequent to the observed loss of polysomes from the cells of the drying gametophytes, i.e., when protein synthesis has ceased. Nevertheless, the transcripts appear to accumulate in the polysomal fraction of cell extracts, suggesting that they are part of a cell component that is large enough to pellet at 100,000g (Fig. 2A,B). Hydrin transcripts do not accumulate in this fashion but remain at a relatively constant level (see Tr416, Fig. 2B). It is unclear at this time whether the accumulation of rehydrin transcripts during the drying phase is a result of selective transcription of rehydrin genes or an increase in rehydrin transcript stability or both. Such an accumulation does not occur if the drying rate is rapid, but rather rehydrin transcripts increase in abundance in the first 90 min following rehydration.

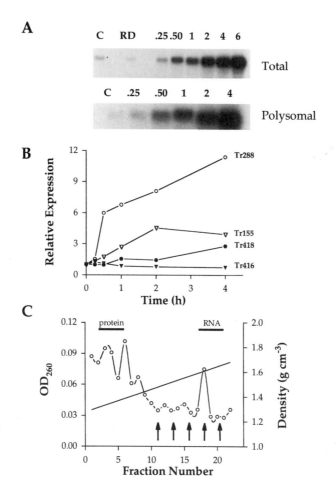

FIGURE 2. (A) Tr288 expression in total and polysomal RNA during slow drying. Total or polysomal RNA was separated by agarose electrophoresis, transferred to nylon membrane and probed with ^{32}P-labeled Tr288 cDNA. C, control 48-hr hydrated moss; RD, rapid-dried moss; numerals represent time points in hours at which gametophytes were removed from a slow-drying sealed chamber (67% relative humidity) described by Oliver (1991).

(B) Relative accumulation of several Tr clones in polysomal pellets during slow drying. Polysomal RNA was treated as described above and probed with ^{32}P-labeled Tr cDNAs. The intensity of hybridization for each clone was quantified, normalized to its control level, and expressed in units of "relative expression." Tr288, Tr155, and Tr418 are rehydrins, Tr416 is a hydrin.

(C) Density analysis of the polysomal pellet from slow-dried *Tortula ruralis*. The 100,000g polysomal pellet was fractionated in a CsCl density gradient (ca. 1.2–1.7 g cm^{-3}). Separation of components within the gradient was monitored by UV absorbance at 260 nm. The solid bars denote the position and densities of pure protein (ca. 1.3 g cm^{-3}) and pure RNA (ca. 1.6 g cm^{-3}). RNA was subsequently purified from each fraction, treated as described above, and probed with ^{32}P-labeled Tr288 cDNA. Arrows denote fractions that hybridized with Tr288.

The observation that the accumulating rehydrin transcripts are recoverable in the polysomal pellets of extracts from slow-dried *T. ruralis* suggested that these mRNAs are sequestered, perhaps in a more stable form, as relatively large conglomerates. A likely candidate for such a structure is the messenger ribonucleo-protein particle (mRNP), a common feature of quiescent eukaryotic cells (see Bag, 1988, for review). mRNPs are cytoplasmic entities (nuclear forms are termed hnRNPs) and are classified into (1) stored mRNPs (e.g., in an untranslated form) or (2) polysomal mRNPs (e.g., a polyribosomal-associated form) (Minich and Ovchinnikov, 1992). The presence of stored, untranslated mRNAs is common in animal embryo development, first reported by Spirin *et al.* (1964), and has been extensively studied in *Xenopus* embryos (reviewed in Davidson, 1986). In plants, there is considerable evidence for the presence of stored mRNAs in the mature, dry seeds of higher plants (Pramanik *et al.*, 1992; Silverstein, 1973). In light of this we have investigated the possibility that the moss prepares for rehydration by storing mRNAs in mRNPs during the drying phase (given that the drying rate is slow) as not only an mRNA protection strategy but also to increase the rapidity of the response once water is available again.

Fractionation of the polysomal pellet from slow-dried *T. ruralis* gametophytes on 10–50% sucrose density gradients and the subsequent analysis of individual fractions for the presence of a rehydrin transcript (Tr288, an abundant rehydrin) revealed that the transcript sedimented at a rate indicating a particle slightly lighter than 40 S ribosomal subunits (Oliver and Bewley, 1997). If the polysomal pellet is isolated under low salt conditions, the transcript is detected in much lower fractions indicating a population of particles of much greater average size (A. J. Wood and M. J. Oliver, unpublished data). In these samples there are no polysomes present (see earlier discussion) and the presence of these larger particles is not discernible by simple OD measurements. A more detailed analysis utilizing cesium chloride gradients demonstrated that the transcript for rehydrin Tr288 can be detected in structures that have a density that is intermediate to pure protein and pure RNA and is less than that reported for ribosomal subunits (Fig. 2C). Such intermediate densities are consistent with the presence of mRNPs and suggests a tight RNA-to-protein interaction (Bag, 1988). It is important to note that the formation of mRNPs in *T. ruralis* during drying requires a sufficient period of time as there is no accumulation of rehydrin mRNAs if water loss is rapid, i.e., to 20% of starting fresh weight in 1 hr.

The formation of mRNPs in response to water loss and their possible roles in mRNA storage and protection have important consequences for the study of vegetative desiccation tolerance and perhaps stress responses of plants in general. The ability to store components during a stress event that are needed for recovery offers a newer dimension to the concept of damage control and the possibility for a more rapid return to growth than does the relatively slower activation and transcription of specific stress or stress-recovery response genes. It is distinctly possible that even in plants where gene activation is the common response to water loss, certain transcripts required for the recovery process are stored in mRNPs during drying. At the least, the determination that the moss *T. ruralis* makes use of stored mRNAs during the rehydration process and beyond should indicate that not all transcripts that are made in response to a stress event are required for immediate use but may be synthesized and stored for when the stress is over and recovery is set in motion.

3.3. Alterations in Transcription

As stated earlier, qualitative alterations in transcription are common in plant responses to stressful environments (Sachs and Ho, 1986) and do occur in the response to desiccation of vegetative tissues of modified desiccation-tolerant plants (Bartels *et al.*, 1990, 1992, 1993; Bartels and Nelson, 1994; Michel *et al.*, 1994; Nelson *et al.*, 1994; Oliver and Bewley, 1997; Piatkowski *et al.*, 1990). The alteration in transcription in modified desiccation-tolerant plants is mediated to a large degree by the plant hormone abscisic acid (ABA), and many of the genes involved share similarity with previously identified ABA responsive genes (Bartels *et al.*, 1990, 1993; Bartels and Nelson, 1994). In the modified desiccation-tolerant angiosperm *Craterostigma plantagineum*, exogenous ABA can induce desiccation tolerance in callous tissue that has lost its ability to withstand drying (Bartels *et al.*, 1990). Similar findings have been reported for the modified desiccation-tolerant moss *Funaria hygrometrica* (Bopp and Werner, 1993; Werner *et al.*, 1991).

In the fully desiccation-tolerant moss *Tortula ruralis*, however, neither a qualitative change in transcription nor the plant hormone ABA appears to play a role in desiccation tolerance. ABA is not detectable in fully hydrated, dry, or rehydrated gametophytes of *T. ruralis* (M. J. Oliver, unpublished data) and we have amply demonstrated that the moss does not exhibit the induction of transcription of genes whose products are novel to the stress event or the recovery from it. Nevertheless, alterations in transcriptional controls have not been precluded, if it is part of the response it is presumably not of major import. Although qualitative changes in transcription have not been detected, changes in the level of transcription of rehydrin genes have been observed (Scott and Oliver, 1994). This is especially true for rehydrins following the rehydration of rapidly dried moss where mRNA accumulation and storage in mRNPs does not occur. Thus, it appears the moss may alter transcriptional controls so as to supplement preexisting pools of rehydrin mRNAs. Whether or not this is a selective change in transcription remains to be determined.

4. CONCLUDING REMARKS

In our introductory remarks we stated the need for simple model plants as tools for the elucidation of the complex mechanisms that exist to enable certain plant species to tolerate extremely stressful environments. The subsequent discussions clearly indicate the value of one set of model plants, the bryophytes (in particular *T. ruralis*), in helping to understand, and to add new dimensions to our understanding of, how plants respond to the extremes of water stress, viz., desiccation. The value of mosses as model plants will increase in the future as we develop the molecular-genetic tools for the manipulation of moss genomes. The haploid genomes of mosses will provide an advantage to more complex plant models, especially in such endeavors as gene tagging, directed mutagenesis, and antisense expression.

Methods already exist for the transformation of at least one bryophyte, the desiccation-intolerant *Physcomitrella patens*, and have been used to great advantage in demonstrating the capability of mosses to properly regulate the transcription of an ABA- and osmotic stress-inducible gene from wheat (Knight *et al.*, 1995). The generation of transgenic gametophytic tissues for desiccation-tolerant species, such as *T. ruralis*, has not been accomplished as yet. The development of procedures to accomplish this will be crucial for the continued utility of moss models for desiccation tolerance as they will be required to identify which of the many gene products involved in the response to desiccation and re-hydration are directly required for desiccation-tolerance. Trans-genic moss tissues will also aid in the assignment of function to such gene products, e.g., rehydrins, that at present have no se-quence similarity to any known genes. At present only three re-hydrins can be assigned a possible function (from sequence anal-ysis); these are an alkyl-hydroperoxidase, a propyl endopeptidase, and polyubiquitin. The confirmation of these possible functions would be greatly advanced by the ability to transfer these coding sequences back into *T. ruralis*, especially as antisense constructs. The development of transformation protocols is a primary goal of all researchers in this area and success is likely given the great strides in such technologies in more complex plants. For the near future, however, the use of mosses as model plants will remain directed toward a more basic understanding of stress-tolerance mechanisms in plants.

Acknowledgments. The authors gratefully acknowledge Brent Mishler at this University and Jepson Herbaria at the University of California, Berkeley for his analysis of the phylogenetic aspects of desiccation tolerance that is included in Fig. 1. We thank John Burke for his helpful comments on the manuscript.

REFERENCES

Alpert, P., and Oechel, W. C., 1987, Comparative patterns of net photosynthesis in an assemblage of mosses with contrasting microdistributions, *Am. J. Bot.* **741**:1787–1796.

Bag, J., 1988, Messenger ribonucleoprotein complexes and translational control of gene expression, *Mol. Genet. (Life Sci. Adv.)* **7**:117–123.

Bartels, D., and Nelson, D., 1994, Approaches to improve stress tolerance using molecular genetics, *Plant Cell Environ.* **17**:659–667.

Bartels, D., Schneider, K., Terstappen, G., Piatkowski, D., and Salamini, F., 1990, Molecular cloning of abscisic acid-modulated genes which are induced during desiccation of the resurrection plant *Craterostigma plantagineum*, *Planta* **181**:27–34.

Bartels, D., Hanke, C., Schneider, K., Michel, D., and Salamini, F., 1992, A desiccation-related Elip-like gene from the resurrection plant *Craterostigma plantagineum* is regulated by light and ABA, *EMBO J.* **11**:277–2778.

Bartels, D., Alexander, R., Schneider, K., Elster, R., Velasco, R., Alamillo, J., Bianchi, G., Nelson, D., and Salamini, F., 1993, Desiccation-related gene products analyzed in a resurrection plant and in barley embryos, in: *Plant Responses to Cellular Dehydration during Environmental Stress* (T. J. Close and E. A. Bray, eds.), American Society of Plant Physiologists, Rockville, MD, pp. 119–127.

Bewley, J. D., 1972, The conservation of polyribosomes in the moss *Tortula ruralis* during total desiccation, *J. Exp. Bot.* **23**:692–698.

Bewley, J. D., 1973a, Polyribosomes conserved during desiccation of the moss *Tortula ruralis* are active, *Plant Physiol.* **51**:285–288.

Bewley, J. D., 1973b, Desiccation and protein synthesis in the moss *Tortula ruralis*, *Can. J. Bot.* **51**:203–206.

Bewley, J. D., 1979, Physiological aspects of desiccation-tolerance, *Annu. Rev. Plant Physiol.* **30**:195–238.

Bewley, J. D., and Krochko, J. E., 1982, Desiccation-tolerance, in: *Encyclopedia of Plant Physiology* (O. L. Lange, P. S. Nobel, C. B. Osmond, and H. Ziegler, eds.), Springer-Verlag, Berlin, Vol. 12B, pp. 325–378.

Bewley, J. D., and Oliver, M. J., 1992, Desiccation-tolerance in vegetative plant tissues and seeds: Protein synthesis in relation to desiccation and a potential role for protection and repair mechanisms, in: *Water and Life: A Comparative Analysis of Water Relationships at the Organismic, Cellular and Molecular Levels* (C. B. Osmond and G. Somero, eds.), Springer-Verlag, Berlin, pp. 141–160.

Bewley, J. D., and Pacey, J., 1978, Desiccation-induced ultrastructural changes in drought-sensitive and drought-tolerant plants, in: *Dry Biological Systems* (J. H. Crowe and J. S. Clegg, eds.), Academic Press, New York, pp. 53–73.

Bewley, J. D., Halmer, P., Krochko, J. E., and Winner, W. E., 1978, Metabolism of a drought-tolerant and a drought-sensitive moss: Respiration, ATP synthesis and carbohydrate status, in: *Dry Biological Systems* (J. H. Crowe and J. S. Clegg, eds.), Academic Press, New York, pp. 185–203.

Bewley, J. D., Reynolds, T. L., and Oliver, M. J., 1993, Evolving Strategies in the adaptation to desiccation, in: *Plant Responses to Cellular Dehydration during Environmental Stress* (T. J. Close and E. A. Bray, eds.), American Society of Plant Physiologists, Rockville, MD, pp. 193–201.

Bliss, R. D., Platt-Aloia, K. A., and Thomson, W. W., 1984, Changes in plasmalemma organization in cowpea radicle during imbibition in water and NaCl solutions, *Plant Cell Environ.* **7**:601–606.

Bopp, M., and Werner, O., 1993, Abscisic acid and desiccation-tolerance in mosses, *Bot. Acta.* **106**:103–106.

Burke, M. J., 1986, The glassy state and survival of anhydrous biological systems, in: *Membranes, Metabolism and Dry Organisms* (A. C. Leopold, ed.), Cornell University Press, Ithaca, NY, pp. 358–363.

Close, T. J., Fenton, R. D., Yang, A., Asghar, R., DeMason, D. A., Crone, D. E., Meyer, N. C., and Moonan, F., 1993, Dehydrin: The protein, in: *Plant Responses to Cellular Dehydration during Environmental Stress* (T. J. Close and E. A. Bray, eds.), American Society of Plant Physiologists, Rockville, MD, pp. 104–118.

Crane, P. R., 1990, The phylogenetic context of microsporogenesis, in: *Microspores: Evolution and Ontogeny* (S. Blackmore and R. B. Knox, eds.), Academic Press, San Diego, pp. 11–41.

Crowe, J. H., Hoekstra, F. A., and Crowe, L. M., 1992, Anhydrobiosis, *Annu. Rev. Physiol.* **54**:579–599.

Davidson, E. H., 1986, *Gene Activity in Early Development*, Academic Press, New York.

Dhindsa, R., 1987, Glutathione status and protein synthesis during drought and subsequent rehydration of *Tortula ruralis, Plant Physiol.* **83**:816–819.

Donoghue, M. J., 1994, Progress and prospects in reconstructing plant phylogeny, *Ann. Mo. Bot. Gard.* **81**:405–418.

Dure, L., III, 1993, A repeating 11-mer amino acid motif and plant desiccation, *Plant J.* **3**:363–369.

Gaff, D., 1980, Protoplasmic tolerance to extreme water stress, in: *Adaptation of Plants to Water and High Temperature Stress* (N. C. Turner and P. J. Kramer, eds.), Wiley–Interscience, New York, pp. 207–230.

Gaff, D., 1989, Responses of desiccation-tolerant "resurrection" plants to water stress, in: *Structural and Functional Responses to Environmental Stresses* (K. H. Krebb, H. Richter, and T. M. Hinkley, eds.), SPB Academic, The Hague, pp. 255–268.

Gaff, D., Bartels, D., Gaff, J. L., and Schneider, K., 1992, Gene expression at low RWC in two hardy tropical grasses, *Trans. Malays. Soc. Plant Physiol.* **3**:238–240.

Gwozdz, E. A., and Bewley, J. D., 1975, Plant desiccation and protein synthesis: An in vitro system from dry and hydrated mosses using endogenous and synthetic messenger RNA, *Plant Physiol.* **55**:340–345.

Gwozdz, E. A., Bewley, J. D., and Tucker, E. B., 1974, Studies on protein synthesis in *Tortula ruralis*: Polyribosome reformation following desiccation, *J. Exp. Bot.* **25**:599–608.

Henckel, R. A., Statrova, N. A., and Shaposnikova, S. V., 1977, Protein synthesis in poikiloxerophyte and wheat embryos during the initial period of swelling, *Sov. Plant Physiol.* **14**:754–762.

Knight, C. D., Sehgal, A., Atwal, K., Wallace, J. C., Cove, D. J., Coates, D., Quatrano, R. S., Bahadur, S., Stockley, P. G., and Cuming, A. C., 1995, Molecular responses to abscisic acid and stress are conserved between moss and cereals, *Plant Cell* **7**:499–506.

Krochko, J. E., Bewley, J. D., and Pacey, J., 1978, The effects of rapid and very slow speeds of drying on the ultrastructure and metabolism of the desiccation-sensitive moss *Cratoneuron filicinum, J. Exp. Bot.* **29**:905–917.

Krochko, J. E., Winner, W. E., and Bewley, J. D., 1979, Respiration in relation to adenosine triphosphate content during desiccation and rehydration of a desiccation-tolerant and a desiccation-intolerant moss, *Plant Physiol.* **64**:13–17.

Leopold, A. C., Bruni, F., and Williams, R. J., 1992, Water in dry organisms, in: *Water and Life. Comparative Analysis of Water Relationships at the Organismic, Cellular and Molecular Levels* (G. N. Somero, C. B. Osmond, and C. L. Bolis, eds.), Springer-Verlag, Berlin, pp. 161–169.

McKersie, B., 1991, The role of oxygen free radicals in mediating freezing and desiccation stress in plants, in: *Active Oxygen and Oxidative Stress and Plant Metabolism* (E. Pell and K. Staffen, eds.), American Society of Plant Physiologists, Rockville, MD, pp. 107–118.

Michel, D., Furini, A., Salamini, F., and Bartels, D., 1994, Structure and regulation of an ABA- and desiccation-responsive gene from the resurrection plant *Craterostigma plantagineum, Plant Mol. Biol.* **24**:549–560.

Minich, W. B., and Ovchinnikov, L. P., 1992, Role of cytoplasmic mRNP proteins in translation, *Biochimie.* **74**:477–483.

Mishler, B. D., Lewis, L. A., Buchheim, M. A., Renzaglia, K. S., Garbary, D. J., Delwiche, C. F., Zechman, F. W., Kantz, T. S., and Chapman, R. L., 1994, Phylogenetic relationships of the "green algae" and "bryophytes," *Ann. Mo. Bot. Gard.* **81**:451–483.

Moore, C. J., Luft, S. E., and Hallam, N. D., 1982, Fine structure and physiology of the desiccation-tolerant mosses, *Barbula torquata* and *Triquetrella papillata* (Mook. F. and Wils.) Broth., during desiccation and rehydration, *Bot. Gaz.* **143**: 358–367.

Nelson, D., Salamini, F., and Bartels, D., 1994, Abscisic acid promotes novel DNA-binding activity to a desiccation-related promoter of *Craterostigma plantagineum*, *Plant J.* **5**:451–458.

Noailles, M. C., 1978, Etude ultrastructurale de la recuperation hydrique apres une periode de secheresse chez une Hypnobryale: *Pleurozium schreberi* (Willd.) Mitt, *Ann. Sci. Nat. Bot.* **19**:249–265.

Oliver, M. J., 1991, Influence of protoplasmic water loss on the control of protein synthesis in the desiccation-tolerant moss *Tortula ruralis*: Ramifications for a repair-based mechanism of desiccation-tolerance, *Plant Physiol.* **97**:1501–1511.

Oliver, M. J., 1996, Desiccation-tolerance in vegetative plant cells, *Physiol. Plant.* **97**:779–787.

Oliver, M. J., and Bewley, J. D., 1984a, Desiccation and ultrastructure in bryophytes, *Adv. Bryol.* **2**:91–131.

Oliver, M. J., and Bewley, J. D., 1984b, Plant desiccation and protein synthesis: IV. RNA synthesis, stability, and recruitment of RNA into protein synthesis upon rehydration of the desiccation-tolerant moss *Tortula ruralis*, *Plant Physiol.* **74**:21–25.

Oliver, M. J., and Bewley, J. D., 1984c, Plant desiccation and protein synthesis: V. Stability of poly(A)$^-$ and poly(A)$^+$ RNA during desiccation and their synthesis upon rehydration in the desiccation-tolerant moss *Tortula ruralis* and the intolerant moss *Cratoneuron filicinum*, *Plant Physiol.* **74**:917–922.

Oliver, M. J., and Bewley, J. D., 1984d, Plant desiccation and protein synthesis: VI. Changes in protein synthesis elicited by desiccation of the moss *Tortula ruralis* are effected at the translational level, *Plant Physiol.* **74**:923–927.

Oliver, M. J., and Bewley, J. D., 1997, Desiccation-tolerance of plant tissues: A mechanistic overview, *Hortic. Rev.* **18**:171–214.

Oliver, M. J., Armstrong, J., and Bewley, J. D., 1993, Desiccation and the control of expression of β-phaseolin in transgenic tobacco seeds, *J. Exp. Bot.* **44**:1239–1244.

Piatkowski, D., Schneider, K., Salamini, F., and Bartels, D., 1990, Characterization of five abscisic acid-responsive cDNA clones from the desiccation-tolerant plant *Craterostigma plantagineum* and their relationship to other water-stress genes, *Plant Physiol.* **94**:1682–1688.

Platt, K. A., Oliver, M. J., and Thomson, W. W., 1994, Membranes and organelles of dehydrated *Selaginella* and *Tortula* retain their normal configuration and structural integrity: Freeze fracture evidence, *Protoplasma* **178**:57–65.

Platt-Aloia, K. A., Lord, E. M., Demason, D. A., and Thomson, W. W., 1986, Freeze-fracture observations on membranes of dry and hydrated pollen from *Collomia, Phoenix* and *Zea, Planta* **168**:291–298.

Pramanik, S. K., Krochko, J. E., and Bewley, J. D., 1992, Distribution of cytosolic mRNAs between polysomal and ribonucleoprotein complex fractions in alfalfa embryos, *Plant Physiol.* **99**:1590–1596.

Reynolds, T. L., and Bewley, J. D., 1993, Characterization of protein synthetic changes in a desiccation-tolerant fern, *Polypodium virginianum.* Comparison of the effects of drying, rehydration and abscisic acid, *J. Exp. Bot.* **44**:921–928.

Sachs, M., and Ho, T. H. D., 1986, Alteration of gene expression during environmental stress, *Annu. Rev. Plant Physiol.* **37**:363–376.

Schonbeck, M. W., and Bewley, J. D., 1981, Responses of the moss *Tortula ruralis* to desiccation treatments. II. Variations in desiccation tolerance, *Can. J. Bot.* **59**:2707–2712.

Scott, H. B., II, and Oliver, M. J., 1994, Accumulation and polysomal recruitment of transcripts in response to desiccation and rehydration of the moss *Tortula ruralis, J. Exp. Bot.* **45**:577–583.

Seel, W. E., Hendry, G. A. F., and Lee, J. E., 1992a, Effects of desiccation on some activated oxygen processing enzymes and anti-oxidants in mosses, *J. Exp. Bot.* **43**:1031–1037.

Seel, W. E., Hendry, G. A. F., and Lee, J. E., 1992b, The combined effects of desiccation and irradiance on mosses from xeric and hydric habitats, *J. Exp. Bot.* **43**:1023–1030.

Sen Gupta, A., 1977, Non-auto-trophic CO_2 fixation by mosses, M.Sc. thesis, University of Calgary.

Siebert, G., Loris, J., Zollner, B., Frenzel, B., and Zahn, R. K., 1976, The conservation of poly (A) containing RNA during the dormant state of the moss *Polytrichum commune, Nucleic Acids Res.* **3**:1997–2003.

Silverstein, E., 1973, Subribosomal ribonucleoprotein particles of developing wheat embryo, *Biochemistry* **12**:951–958.

Simon, E. W., 1978, Membranes in dry and imbibing seeds, in: *Dry Biological Systems* (J. H. Crowe and J. S. Clegg, eds.), Academic Press, New York, pp. 205–224.

Simon, E. W., and Mills, L. K., 1983, Imbibition, leakage, and membranes, in: *Mobilization of Reserves in Germination* (C. Nozzolillo, P. J. Lee, and F. A. Loewus, eds.), Plenum Press, New York, pp. 9–27.

Smirnoff, N., 1993, The role of active oxygen in the response of plants to water deficit and desiccation, Tansley Review No 52, *New Phytol.* **125**:27–58.

Smith, I. K., Polle, A., and Rennenberg, H., 1990. Glutathione, in: *Stress Responses in Plants: Adaptation and Acclimation Mechanisms* (R. G. Alscher and J. R. Cummings, eds.), Wiley–Liss New York, pp. 201–215.

Spirin, A. S., Belitsina, N. V., and Ajtkhozhin, M. A., 1964, Messenger RNA in early embryogenesis, *Zh. Obshch. Biol.* **25**:321–338.

Stewart, G. R., and Lee, J. A., 1972, Desiccation-injury in mosses. II. The effect of moisture stress on enzyme levels, *New Phytol.* **71**:461–466.

Stewart, R. R. C., and Bewley, J. D., 1982, Stability and synthesis of phospholipids during desiccation and rehydration of a desiccation-tolerant and a desiccation-intolerant moss, *Plant Physiol.* **69**:724–727.

Swanson, E. S., Anderson, N. H., Gellerman, J. L., and Schlenk, H., 1976, Ultrastructure and lipid composition of mosses, *Bryologist,* **79:**339–349.

Thomson, W. W., and Platt-Aloia, K. A., 1982, Ultrastructure and membrane permeability in cowpea seeds, *Plant Cell Environ.* **5:**367–373.

Tucker, E. B., and Bewley, J. D., 1976, Plant desiccation and protein synthesis. III.Stability of cytoplasmic RNA during dehydration and its synthesis on rehydration of the moss *Tortula ruralis, Plant Physiol.* **57:**564–567.

Tucker, E. B., Costerton, J. W., and Bewley, J. D., 1975, The ultrastructure of the moss *Tortula ruralis* on recovery from desiccation, *Can. J. Bot.* **53:**94–101.

Werner, O., Espin, R. M. R., Bopp, M., and Atzorn, R., 1991, Abscisic-acid-induced drought tolerance in *Funaria hygrometrica* Hedw., *Planta* **186:**99–103.

2

Local and Systemic Plant Defensive Responses to Infection

R. HAMMERSCHMIDT

1. INTRODUCTION

Plants are subject to infection by a wide range of pathogen types that include fungi, protists, bacteria, viruses, and nematodes (Agrios, 1988). Although many thousands of plant pathogens exist in nature, only a few are capable of successfully infecting and establishing a disease-causing compatible (susceptible) interaction with any specific type of plant. However, in many cases where the compatible interaction is established, devastating epidemics resulting in large losses of plants and plant products can develop. One only needs to consider the potato blight that caused serious losses and famine in Ireland and Europe in the 1840s to realize how severe plant disease can become.

Plant pathogens often show great specificity for the host plants that they can successfully infect and parasitize (Heath, 1994). Although some pathogens like the fungus *Sclerotinia scle-*

R. HAMMERSCHMIDT • Department of Botany and Plant Pathology, Michigan State University, East Lansing, Michigan 48824.

Stress-Inducible Processes in Higher Eukaryotic Cells, edited by Koval. Plenum Press, New York, 1997.

rotiorum have a very wide host range, most pathogens are restricted to one or a few host species (Agrios, 1988). Furthermore,
within a particular species that can be infected, the pathogen may
only be able to infect a particular cultivar or even a specific plant
part or developmental stage of a plant (Heath, 1995). The high level
of specificity that most pathogens may have for a single or limited
number of host plant species indicates that plants are resistant to
the vast majority of pathogens with which they may come into
contact.

Understanding the mechanisms of plant disease resistance
has the potential for developing new strategies for the control of
plant disease that is safe and effective. Because plants are resistant to most pathogens, it is logical to assume that plants have all
necessary genetic information to defend themselves against most
if not all pathogens if these "defense" genes can be activated soon
enough after infection to be effective.

2. TYPES OF DISEASE RESISTANCE

Heath (1995) has classified plant disease resistance into several major categories. These are: "basic resistance," which is the
type of resistance expressed by a plant species to the pathogens of
other plants that cannot cause disease on this species; "parasite-
specific resistance," which is the resistance expressed by cultivars
within a plant species that is a susceptible host for the pathogen;
"age-related resistance" in which the age of the plant tissue or part
affects the resistance or susceptibility to the pathogen; "organ-
specific resistance" in which not all parts or tissues of a plant are
susceptible to infection; and "induced resistance" in which resistance is expressed in a susceptible plant after a prior infection or
treatment with a resistance inducing agent. Each of these types of
resistance has unique properties based on the genetic and/or
physiological state of the plant. However, investigations of the
cellular mechanisms used to express each type of resistance suggest that the same basic set of defense-related genes are used in
many facets of plant defense irrespective of the type of resistance
(Heath, 1994). Based on observations of these types of resistance,
it is likely that one of the differences between resistance and
susceptibility of a plant to infection may be based on how quickly
the genes that control defense are activated (Kuć, 1995a). Of these

five types of resistance, parasite-specific and induced resistance have received the most study regarding mechanisms of defense and the cellular responses that control these mechanisms. Because of this, the expression of defense in these two types of resistance will be the focus of this review.

2.1. Parasite-Specific Resistance

Parasite-specific resistance typically refers to resistance that is controlled by a single resistance gene in a host plant, and that some members of the plant species are a host for the pathogen (Heath, 1995; Staskawicz *et al.*, 1995). One type of parasite-specific resistance is based on the ability of the host plant to detoxify toxins produced by the pathogen. For example, resistance of maize to *Cochliobolus carbonum* is based on the ability of the resistant plant to detoxify the toxin produced by this fungus (Meeley *et al.*, 1992). However, resistance is generally thought to be controlled by the interaction of a resistance (R) gene in the host plant and a corresponding avirulence (avr) gene in the pathogen (Heath, 1994). This type of genetic interaction between host and pathogen is often referred to as a "gene-for-gene" interaction.

"Gene-for-gene"-based resistance dictates that only in cases where the plant contains an R gene whose product can specifically recognize the product of the pathogen avirulence gene will the plant resistance response be expressed (Staskawicz *et al.*, 1995; Heath, 1994). On the contrary, if the plant has an R gene but the isolate of the pathogen does not have the corresponding avr gene, then the interaction will be compatible (susceptible). Similarly, if the pathogen has an avr gene but the host does not have the corresponding R gene, a compatible state will also develop on infection.

The function of R and avr genes is now being elucidated by the cloning and sequencing of these genes. Resistance genes have been cloned and sequenced from a number of plant species (Michelmore, 1995; Staskawicz *et al.*, 1995). These include Arabidopsis, tomato, tobacco, and flax. Although these R genes are specific for resistance to bacteria (Arabidopsis and tomato), fungi (tomato and flax), and a virus (tobacco), many have similar characteristics. For example, the flax *L6* gene, the tobacco *N* gene, the tomato *Cf-9* gene, and the Arabidopsis *RPS2* gene all code for proteins with leucine-rich repeats, and sequence analysis has suggested that these proteins may be important in recognition (Staskawicz et al., 1995).

Many of the pathogen avr genes that correspond to specific plant R genes have also been cloned and sequenced (Leach and White, 1996; Michelmore, 1995). Although the exact function of most of these genes is not known, one avr gene product is a peptide secreted by the fungal pathogen of tomato *Cladosporium fulvum* that appears to be directly involved in eliciting the hypersensitive defense response in a resistant plant (Van Kan et al., 1991), and another is involved in biosynthesis by *Pseudomonas syringae* of a low-molecular-weight elicitor, called a syringolide, that induces the host cell resistance response (Smith et al., 1993). The other avr genes that have been cloned from bacteria all appear to code for hydrophilic proteins with no known or recognizable function (Leach and White, 1996).

How the R and avr gene products function in the expression of the resistance in plants with the appropriate R genes is an area currently under investigation. It has been hypothesized, based on sequence analysis, that the R gene product serves as a receptor or other protein involved in signal transduction (Staskawicz et al., 1995). Using both transient expression of an avr gene in plant cells and a yeast–two hybrid system, two research groups have recently provided evidence that the product of the tomato R gene *Pto* physically interacts with the product of the avirulence gene coded by the *Pseudomonas syringae* pv. *tomato* avirulence gene *avrPto* (Scofield et al., 1996; Tang et al., 1996). The specific interaction between the avr gene product and the R gene product then results in a cascade of events that lead to the expression of defense responses (Lamb, 1996), some of which will be discussed later.

2.2. Induced Resistance

In the previous section, resistance that is based on specific gene-for-gene interactions that lead to the expression of defense genes was described. Another type of resistance observed in plants is not based on specific recognition, but rather is a response to a previous infection or certain chemical treatments that results in expression of resistance to a wide variety of diseases in plants that are classified as susceptible to some or all pathogens. This type of resistance is known as induced or acquired resistance (Hammerschmidt and Becker, 1997; Hammerschmidt and Kuć, 1995; Ryals et al., 1994). Induced resistance is generally activated by a prior localized infection that results in a necrotic lesion. The induced resistance can be localized near the inducing treatment or can become systemic throughout the plant. Thus, this phenomenon is superficially similar to immunization in animals. Induced resistance, and specifically systemic induced resistance, has been described for a number of plant species across several flowering plant families (Hammerschmidt and Kuć, 1995). In almost all cases, the resistance is induced by localized infection of one leaf or one plant part with a pathogen that elicits a localized, necrotic lesion (Hammerschmidt, 1993). The end result is a plant that is now resistant not only to the initial inducing pathogen, but also to a wide variety of fungal, bacterial, and viral pathogens. Thus, induced resistance, as compared with parasite-specific resistance, is very parasite-nonspecific.

The induction of systemic resistance can be divided into four phases (Ryals et al., 1994; Hammerschmidt, 1993). Induced resistance starts with the initial interaction of the plant with a pathogen that results in the generation of a primary resistance signal; the primary signal is then translocated throughout the plant; the primary signal, or secondary signals induced by the primary signal, then activate a set of defense-associated genes throughout the plant body; finally, the plant now has a heightened ability to express additional defenses when subsequently attacked by a pathogen. The nature of the signals that are involved in resistance expression will be discussed later. The type of systemic gene expression that is associated with induced resistance is generally reflected in the new transcription and translation of a set of proteins called PR or pathogenesis-related proteins as well as oxidases such as peroxidase, phenoloxidase, and lipoxygenase (Stermer, 1995). When a plant exhibiting induced resistance is subsequently inoculated with a pathogen, the infected cells rapidly respond with the induction of further defenses, many of which, if not all, are utilized in defense triggered by the interaction of R gene and avr gene products. The nature of some of these types of defense will be described below.

3. TYPES OF DEFENSE RESPONSES

The expression of resistance to an invading pathogen is generally thought to require the presence of constitutive defenses and/or the activation of a variety of defense responses at the cellular level (Hammerschmidt and Schultz, 1996). Whether the type of resistance is based on a gene-for-gene interaction (i.e., parasite-specific resistance) or induced resistance, mechanisms must be put in place to stop the invading pathogen. Although there are few examples of preformed or constitutive defenses playing a decisive role in resistance [e.g., presence of a group of saponins called avenacins in oats that confer resistance and dictate host range to the fungal pathogen *Gaumanomyces graminis* (Bowyer *et al.*, 1995; Osbourn *et al.*, 1994)], most resistance is expressed in the form of multiple mechanisms elicited as a result of infection (Hammerschmidt and Schultz, 1996). This section will discuss several of these active responses that are utilized in the expression of parasite-specific and induced resistance. Several recent reviews cover most of these defenses in detail, and the reader is directed to these for further information (Baker and Orlandi, 1995; Kuć, 1995b; Dixon *et al.*, 1994; Goodman and Novacki, 1994; Nicholson and Hammerschmidt, 1992).

3.1. The Hypersensitive Response

The terms *hypersensitive response* or *hypersensitivity* are used in plant pathology to describe the localized and rapid death of one or a few host plant cells in response to invasion by an avirulent isolate of a pathogen (Goodman and Novacki, 1994). The hypersensitive death of host cells is characterized by a rapid decompartmentalization of the internal structure of the cell, loss of membrane integrity, and the appearance of yellow to brown phenolic oxidation products (Goodman and Novacki, 1994). The hypersensitive response alone may be an effective defense against those pathogens that are obligately parasitic on plants (e.g., rusts, powdery mildews, some downy mildews), but this defense response appears to be only one part of a large arsenal of defenses that plants use to combat infection.

The hypersensitive response of the plant cell to the pathogens appears to be coordinately expressed with other defenses (Goodman and Novacki, 1994). Because of this, it has been suggested that the hypersensitive response is a type of programmed cell death (Greenberg *et al.*, 1994). Although the phenomenon of programmed cell death has been studied extensively in animal systems, similar studies have only recently begun in plants. For example, the differentiation of xylem elements that are dead on functional maturity may be an example of this phenomenon (Mittler and Lam, 1995). Several mutants of plants have been isolated that express the spontaneous development of necrotic lesions that resemble hypersensitive necrosis (Dietrich *et al.*, 1994; Greenberg *et al.*, 1994) . These plants also express many of the genes and defenses that are typical of pathogen resistance as the cell death occurs (Greenberg *et al.*, 1994). Because of the similarity of these spontaneous lesions to the hypersensitive response, it is hoped that genetic and molecular analysis of these mutants may shed some light on how hypersensitive host cell death may be regulated or programmed and how this process of cellular "suicide" is coordinated with resistance expression.

Pathogen-induced cell death that has characteristics of programmed cell death or apoptosis in animal cells has recently been reported. Ryerson and Heath (1996), in studies on the hypersensitive response of cowpea (*Vigna unguiculata*) cells to incompatible races of the rust fungus *Uromyces vignae*, showed that two phenomena observed in mammalian apoptosis could also be observed in the plant hypersensitive response. The dying host cells had typical DNA laddering. They also reported that cytological studies of the hypersensitively responding cells using the TUNEL procedure showed *in situ* fragmentation of the nuclear DNA. The response was not, however, completely specific to the pathogen-induced hypersensitive response. Treatment of tissues with KCN induced a similar apoptosislike response. However, other chemical treatments that killed cells did not induce DNA laddering or cytologically detectable nuclear DNA fragmentation. Levine *et al.* (1996) reported that Arabidopsis leaves and suspension-cultured cells of tobacco and soybean also underwent an apoptosislike cell death in response to treatments with avirulent pathogens or pathogen-produced inducers of hypersensitive cell death.

Although good evidence for programmed cell death in a resistance response to pathogens has been presented, the response is not restricted to disease interactions that lead to resistance. Wang *et al.* (1996) found that treatment of tomato tissue with a fungal toxin that selectively kills cells of certain tomato cultivars will also induce an apoptosislike response. This is interesting because the fungal toxin used in this study is required for disease development by the pathogen. Thus, it seems very likely that programmed cell death in plants in response to pathogen-associated phenomena is not restricted to resistance-associated responses.

3.2. Increases in Activated Oxygen and Oxidative Enzyme Activity

The observation that cells undergoing the hypersensitive response turn brown as endogenous phenols begin to oxidize suggests that there is the production of activated oxygen species (Goodman and Novacki, 1994). Pathogen induced production of activated oxygen species as part of active defense has received considerable attention in the last few years and has been reviewed (Baker and Orlandi, 1995; Mehdy, 1994; Sutherland, 1991). The types of oxygen species that are known to be produced during host cell resistance responses include hydrogen peroxide, hydroxyl radical, and superoxide anion. Each of these active oxygen species has the potential of being cytotoxic (Baker and Orlandi, 1995) as well as assisting in cross-linking reactions that strengthen the plant cell wall (Brisson et al., 1994).

It has also been suggested that hydrogen peroxide may function in the defense signaling pathway. Levine et al. (1994) have shown that the production of hydrogen peroxide during the "oxidative burst" exhibited by plant cells during the early phases of defense reactions may act as a signal that activates defense genes in surrounding cells.

Along with the appearance of the activated oxygen species is the activation and/or de novo synthesis of three types of oxidases: peroxidase, phenoloxidase, and lipoxygenase. Increase in peroxidase activity is a common response to infection (Smith et al., 1991; Smith and Hammerschmidt, 1988) and appears to rely on new transcription (Rasmussen et al., 1995). Peroxidase may be involved in generation of antimicrobial levels of hydrogen peroxide (Peng and Kuć, 1992) as well as in strengthening the cell wall via lignin polymerization or the formation of phenolic cross-links (Nicholson and Hammerschmidt, 1992). Phenoloxidases may contribute to defense via the production of toxic quinones (Appel, 1993), and the activation of this enzymatic activity is associated with early events of the hypersensitive response (Lazarovits and Ward, 1982). Lipoxygenase may contribute to the hypersensitive response via disruption of cell membrane lipids (Goodman and Novacki, 1994), to defense through the formation of toxic lipid oxidation products (Croft *et al.*, 1993), and to the synthesis of lipid-derived signal molecules like jasmonic acid (Farmer, 1994).

3.3. Cell Wall Modifications

During the infection process and subsequent development of disease, most pathogens must penetrate through cell walls or the pectin-rich middle lamella that is found in between adjacent cells (Schäfer, 1994; Aist, 1983). Although the native cell wall may serve as a barrier to some pathogens, most pathogens have effective means of penetrating plant cell walls (Schäfer, 1994).

Infection of plants by many fungi requires the formation of specialized infection structures called appressoria from which the fungus will directly penetrate through the outer epidermal cell wall or penetrate through stomatal openings (Schäfer, 1994). The appressorium firmly attaches to the cuticle of the epidermis and at the base of the appressoria a thin strand of hyphae called a penetration peg is produced. The penetration peg then penetrates the cuticle and underlying polysaccharide cell wall materials by enzymatic and/or mechanical means. Once the pathogen has passed the outer cell wall, it is further presumed that a battery of pectolytic and cellulolytic enzymes are used as the pathogen begins to colonize the tissue (Schäfer, 1994).

Penetration of plant cell walls is not generally very difficult for pathogens. However, one of the types of defense responses observed in resistant interactions is a modification of the cell wall with structural components that increase the mechanical strength of the wall or decrease the susceptibility of the cell wall to pathogen-produced cell-wall-degrading enzymes (Ride, 1978).

One type of wall modification that can accomplish both of these features is the deposition of the phenolic polymer lignin (Nicholson and Hammerschmidt, 1992). Lignin is synthesized by the free-radical-mediated polymerization of hydroxycinnamyl alcohols that are derived from phenylalanine. During the early stages of infection into a resistant plant, the invaded cell often deposits lignin in its cell wall (e.g., Hammerschmidt *et al.*, 1985). In some cases, that lignification occurs after the cell wall has been penetrated, and thus the cell wall changes may function to contain the pathogen in one place until other defenses (e.g., accumulation of phytoalexins or hydrolytic enzymes) are induced (Ride, 1978). In other cases, the deposition of lignin may rapidly occur in the outer epidermal cell wall at the point of attempted fungal penetration and thus may function to prevent the pathogen from penetrating into the cell (Hammerschmidt *et al.*, 1985; Aist, 1983).

It has been demonstrated that expression of nonhost (Hammerschmidt *et al.*, 1985), parasite-specific (Hammerschmidt and Kuć, 1982), and induced resistance (Stein *et al.*, 1993; Hammerschmidt and Kuć, 1982) in cucumber is, at least partly, the result of rapid lignin deposition at the point of attempted penetration by fungi into epidermal cells. The response is highly localized and often only a very small part of the epidermal cell wall is involved. Interestingly, if the pathogen does successfully invade the cell, the entire epidermal cell often will lignify (Hammerschmidt *et al.*, 1985), and ligninlike polymers are even deposited on the growing tip of fungal hyphae in the host cell (Stein *et al.*, 1993). Similar types of responses have been seen in other plant species (Mauch-Mani and Slusarenko, 1996).

The second-to-last enzymatic step in lignin biosynthesis is the reduction of a hydroxycinnamaldehyde to the corresponding alcohol by cinnamyl alcohol dehydrogenase (CAD) (Moersbacher *et al.*, 1990). Several reports show that CAD-specific inhibitors will block the synthesis and deposition of lignin and the expression of parasite-specific resistance (Zeyen *et al.*, 1995; Moersbacher *et al.*, 1990).

CAD inhibitors have also been reported to block the hypersensitive response (Zeyen *et al.*, 1995; Moersbacher *et al.*, 1990). Lignification utilizes hydroxycinnamyl alcohols and hydrogen peroxide that react together in the presence of peroxidase to form reactive free radicals of the hydroxycinnamyl alcohols. Thus, the production of the hydroxycinnamyl alcohols and the free radicals produced from them may be an important aspect of the expression of the hypersensitive response. This observation, along with observations that programmed cell death plays a role in the maturation of the lignified xylem elements (Mittler and Lam, 1995), suggests that lignification of cells may be an integral part of the programmed death hypothesized to be important in hypersensitive cell death.

3.4. Phytoalexins

Most flowering plant families respond to infection with the synthesis of low-molecular-weight antibiotics known as phytoalexins (Kuć, 1995b). The original concept of phytoalexins was of compounds that were specifically involved in host resistance. However, phytoalexins are now generally defined more broadly as compounds that are antimicrobial and induced after infection (Paxton, 1981).

Phytoalexins represent a diverse group of compounds from a number of different secondary metabolic pathways (Kuć, 1995b). For example, plants in the Solanaceae produce sesquiterpenoid phytoalexins via the mevalonate pathway whereas legumes produce isoflavonoid or pterocarpan-derived phytoalexins that are synthesized using both the shikimate pathway and the acetate–malonate pathway. Several examples of phytoalexins representing a diversity of biosynthetic origins are shown in Fig. 1. Although the chemistry of phytoalexins differs from one plant family to another, it can be generalized that an incompatible interaction between a pathogen and host plant results in the rapid *de novo* synthesis and accumulation of phytoalexins (Kuć, 1995b). Many studies have shown that the inhibition of pathogen development is often correlated with the accumulation of phytoalexins at the site of infection (e.g., Pierce *et al.*, 1996; Snyder and Nicholson, 1990). Genetic analysis has also shown cosegregation of resistance with pathogen-induced phytoalexin accumulation in oat lines resistant to the crown rust pathogen, *Puccinia coronata* (Mayama *et al.*, 1995).

Concomitant with the synthesis and accumulation of phytoalexins is the induction of new transcription and translation of genes that are involved in the biosynthesis of these compounds. For example, the infection of legumes with incompatible pathogens or treatment of tissues with molecules known to elicit phytoalexin synthesis will result in the rapid transcription and translation of genes that are involved in the phenylpropanoid and flavonoid pathways (e.g., phenylalanine ammonia lyase, chalcone synthase), whereas in solanaceous plants, genes regulating terpene synthesis (e.g., HMGR CoA reductase and sesquiterpene cyclases) are induced (Kuć, 1995b).

Unfortunately, most of the evidence for phytoalexins in defense is correlative and definitive evidence has not been readily forthcoming to support a role for these compounds in defense. However, there are several examples where a case can be made for a role in defense. *In situ* analysis of phytoalexin concentrations in relation to fungal or bacterial pathogen development has

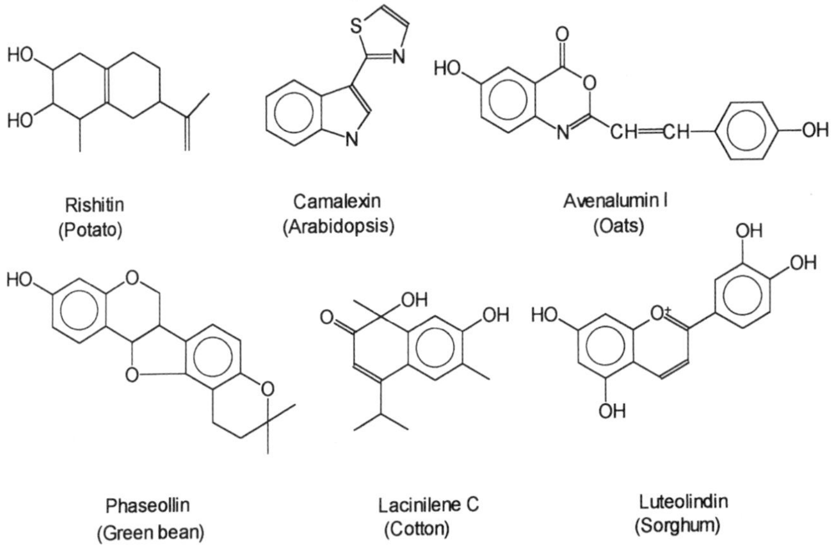

FIGURE 1. Examples of phytoalexin structures. Rishitin and lacinoline C are derived from the acetate–mevalonate pathway and are examples of sesquiterpenoid phytoalexins. Phaseolin and luteolindin are pterocapan and flavonoid phytoalexins that are derived from joint activities of the shikimic acid pathway and acetate–malonate pathway. Avenalumin I is derived from the shikimic acid pathway as is the indole moiety of camalexin.

also been used to support a role for phytoalexins in resistance. Nicholson and co-workers have carried out a series of microspectrophotometric measurements of the accumulation of deoxyanthoxyanidin (e.g., luteolindin; Fig. 1) phytoalexins in the epidermal cells of sorghum plants undergoing a resistant reaction to incompatible fungi (Snyder et al., 1991; Snyder and Nicholson, 1990). These compounds absorb visible light (red to orange in color), and thus are easily observed in fresh tissue. It was found that as the pathogen attempts to infect, the host cell begins to produce small inclusion bodies that appear to be the site of synthesis of these phytoalexins. The inclusions gradually migrate to the site of the infecting hyphae and, as they migrate, change in color from clear to red-orange (indicating that synthesis of the final compounds has occurred). Microspectrophotometric analysis of these inclusion-containing cells indicated that the concentration of phyto-alexin in these cells exceeds what is needed to kill the pathogen *in vitro* (Snyder *et al.*, 1991). This response does not occur in suscept-ible interactions, and thus is a very specific response of the resis-tant host cells to infection.

Similar types of experiments using microspectrofluorometry have shown that the accumulation of the fluorescent phytoalexins of oats (avenalumins, Fig. 1) and cotton (lacinilene C, Fig. 1) also correlate very well in location, timing, and concentration with the cessation of pathogen growth in resistant plants (Pierce *et al.*, 1996; Essenberg *et al.*, 1992; Mayama and Tani, 1982). Unfortu-nately, most phytoalexins are not colored or fluorescent and can-not be studied in this manner.

Mutants that are deficient in phytoalexin production provide another strategy to study the role of these compounds in plant defense. The small crucifer Arabidopsis produces an indole-based phytoalexin with the trivial name camalexin (Tsuji *et al.*, 1992; Fig. 1) and the accumulation of this compound in response to incompatible bacterial and fungal pathogens suggests that it may play a role in resistance (Zook and Hammerschmidt, 1997; Tsuji *et al.*, 1992). Glazebrook and Ausubel (1994) have isolated a series of mutants that are either reduced or totally deficient in the produc-tion of camalexin. Although not any more susceptible to infection by incompatible bacterial pathogens, there is an effect on the multiplication of virulent bacterial pathogens. Further analysis of these mutants will be of value in determining the relative role of camalexin in plant resistance. Parallel studies on the biosynthesis of camalexin (Zook and Hammerschmidt, 1997) suggest that few biosynthetic genes may be specifically needed for synthesis. Thus, it should be possible to determine which genes are required for biosynthesis, and then use these genes in definitive studies on the role of camalexin in defense.

3.5. Pathogenesis-Related Proteins

A specific group of proteins associated with the expression of resistance are known as the pathogenesis-related (PR) proteins. The PR proteins have been classified into subgroups PR1 through PR11 (Van Loon *et al.*, 1994). Several comprehensive reviews on the regulation and role of PR proteins have been published (e.g.,

Cutt and Klessig, 1992; Linthorst, 1991). Thus, only a brief summary of some of these proteins and the evidence supporting their role in resistance will be described in this section.

The role that PR proteins play in resistance is based on several lines of evidence. First, these proteins are both locally induced during the expression of the hypersensitive response (Linthorst, 1991) and they are also systemically expressed in plants exhibiting systemic induced resistance (Ward et al., 1991). Second, β-1,3-glucanases (PR2) and chitinases (PR3) may be involved in resistance by their action on fungal cell walls (Stermer, 1995; Linthorst, 1991). Finally, transformation of plants with PR genes has also provided a means to test the role of these proteins in resistance. For example, the constitutive expression of PR1 in transgenic tobacco or PR5 in transgenic potato increases resistance to oomycete pathogens (Liu et al., 1994; Alexander et al., 1993), and this resistance may be related to the observations that PR proteins exhibit antimicrobial activity in vitro (Abad et al., 1996; Enkerli et al., 1993).

4. ENDOGENOUS SIGNALS FOR RESISTANCE

Expressions of resistance in both parasite specific and induced resistance often share similar types of mechanisms (Heath, 1994). Because of this, it is likely that some common signals are shared in the expression of defense genes in these two types of resistance. There have been several compounds proposed that could function as local or systemic signals for resistance expression. These include oligogalacturonides derived from the cell walls of plants, jasmonic acid derived from membrane linolenic acid, peptides, and the simple phenol salicylic acid (Farmer 1994; Klessig and Malamy, 1994; Pierpont, 1994; Hammerschmidt, 1993). Of these putative signals, salicylic acid (Fig. 2) is the most likely candidate for one of the endogenous disease resistance signals.

4.1. Salicylic Acid Is a Signal for Resistance

White (1979) first reported that treating tobacco plants with salicylic acid enhanced the resistance of the plants to infection by tobacco mosaic virus. This was supported by the observations of Mills and Wood (1984) and Rasmussen et al. (1991) who demon-

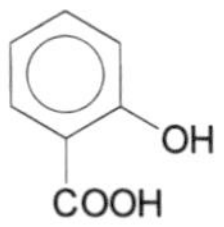

Salicylic Acid

FIGURE 2. Structure of salicylic acid (2-hydroxybenzoic acid).

strated that cucumber could be protected against infection by *C. lagenarium* by prior treatment with salicylic acid. The resistance-inducing effects of salicylic acid have been documented for several other host–pathogen systems (see review by Pierpont,1994). These observations, along with the fact that salicylic acid will induce the expression of PR proteins and that it is a naturally occurring plant product (Pierpont, 1994), suggest that this hydroxybenzoic acid could function as an endogenous signal in both locally expressed as well as systemically induced resistance responses.

4.2. Biosynthesis of Salicylic Acid

One of the biochemical changes that almost always occurs in infected plants is the induction of phenolic compound biosynthesis from phenylalanine (Nicholson and Hammerschmidt, 1992). Because of this, it is likely that infection will promote the synthesis of salicylic acid. Raskin and co-workers have recently shown that phenylalanine is the initial precursor for salicylic acid (Leon *et al.*, 1993, 1995a; Yalpani *et al.*, 1993). They found that cinnamic acid, the product of phenylalanine ammonia lyase (PAL) action on phenylalanine, is converted to benzoic acid. Benzoic acid is then converted to salicylic acid by the enzyme benzoic acid 2-hydroxylase (Leon *et al.*, 1995b). The general pathway is shown in Fig. 3. Knowing which enzymes are involved in salicylic acid biosynthesis will provide another framework for studies on the role of the phenol in resistance.

4.3. Role of Salicylic Acid in Local Resistance

The resistance-inducing properties of exogenous salicylic acid and the synthesis of this compound in infected plants suggest that salicylic acid may play a role in host resistance. Excellent evidence for a role for endogenous salicylic acid in the expression

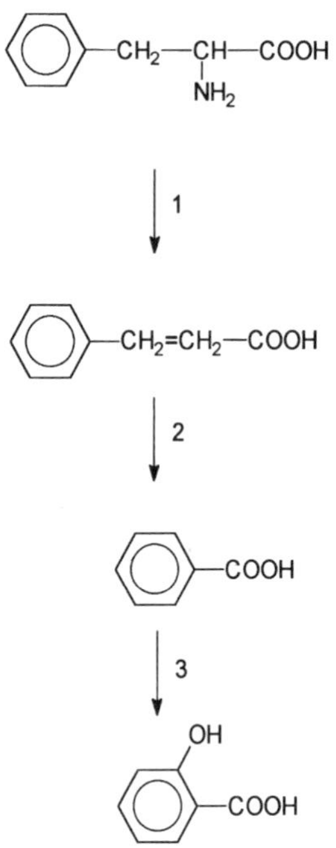

FIGURE 3. Biosynthesis of salicylic acid. Step 1 is the conversion of phenylalanine to cinnamic acid by phenylalanine ammonia lyase, step 2 is the oxidation of cinnamic acid to benzoic acid (enzymatic mechanism not determined), and step 3 is the conversion to benzoic acid to salicylic acid by benzoic acid 2-hydroxylase.

of resistance was reported by Gaffney *et al.* (1993). They transformed tobacco with the bacterial *nahG* gene that codes for the enzyme salicylate hydroxylase. This enzyme converts salicylic acid to catechol, which they found has no disease resistance-inducing activity. The transformed plants expressing salicylate hydroxylase did not accumulate salicylic acid in response to tobacco mosaic virus (TMV) infection. Interestingly, TMV lesions formed on the *nahG* transformants were larger than on the wild-type controls. Tobacco and Arabidopsis transformed with *nahG* were later shown to be more susceptible to other pathogens (Delaney *et al.*. 1994). As salicylic acid is not very toxic to most pathogens (e.g., Mills and Wood, 1984), it seems likely that the reason for the enhanced susceptibility was the lack of salicylic acid-mediated defense induction. Maher *et al.* (1994) reported that tobacco plants transformed to suppress the activity of phenylalanine ammonia lyase were more susceptible to fungal infection. Although the initial explanation was that the plants were more susceptible because of a lack of induced phenolic defenses, a more recent report suggests that the plants were more susceptible because they could not produce sufficient quantities of salicylic acid to be used in regulating the defense response (Pallas *et al.*, 1996). Using Arabidopsis, Mauch-Mani and Slusarenko (1996) provided further evidence that a major role for induced PAL activity in infected plant tissues was for the production of salicylic acid.

4.4. Salicylic Acid and Systemic Resistance

The ability of salicylic acid to induce defense associated gene expression and resistance when applied to plant tissues and the compound's natural occurrence in plants (Pierpont, 1994) suggested that it might function as the endogenous, translocated signal for systemic induced resistance that had been predicted to exist (e.g., Dean and Kuć, 1986a,b). Infection of tobacco and cucumber on one leaf to induce resistance resulted in systemic increases in salicylic acid content during the onset of systemic induced resistance (Malamy *et al.*, 1990; Métraux *et al.*, 1990). In the work reported by Métraux *et al.* (1990), the increase in salicylic acid was measured in phloem exudates, thus supporting the hypothesis that salicylic acid is the translocated signal. Following those reports, a number of papers were published that supported a role for salicylic acid in the induced resistance phenomenon (for review see Pierpont, 1994).

Despite a large body of literature supporting a role for salicylic acid in resistance induction, these papers did not directly address the question of salicylic acid as the translocated signal. For salicylic acid to be the translocated signal, it must be produced in the infected leaves and then transported (in sufficient quantities) to account for the increases in salicylic acid observed throughout the plant. Rasmussen *et al.* (1991) tested this hypothesis using the cucumber–*Pseudomonas syringae* pv. *syringae* system. Smith *et al.* (1991) had previously demonstrated that infection of cucumber with this bacterium would result in expression of systemic resistance within 24 hr. They also showed that the signal for systemic resistance was generated within 4 to 6 hr after inoculation with *P. syringae*. Using the time course of resistance induction established by Smith *et al.* (1991), Rasmussen *et al.* (1991) detached *Pseudomonas*-inoculated leaves of cucumber at intervals after inoculation in order to determine the time of salicylic acid transport out of the infected leaf. Salicylic acid could be detected in the phloem exudates of the *Pseudomonas*-inoculated leaf by 8 hr after inoculation, and increases in salicylic acid occurred in the phloem of the leaf above the inoculated leaf by 12 hr. These data suggested that salicylic acid might be moving from one leaf to another. However, in plants that had the inoculated leaf detached before there was a detectable increase in salicylic acid, the entire plant contained greatly increased levels of salicylic acid when analyzed 24 hr after the initial inoculation. Based on these results, Rasmussen *et al.* (1991) concluded that salicylic acid was a secondary signal that was induced by a primary, translocated signal generated at the infection site. The systemic biosynthesis of salicylic acid in infected cucumber plants has been found through the *in situ* labeling of salicylic acid (Meuwly *et al.*, 1995). Thus, there is further evidence that salicylic acid is synthesized throughout the plant in response to a primary signal.

The conclusion of Rasmussen *et al.* (1991), that salicylic acid was not the translocated resistance signal, was confirmed in tobacco by Vernooij *et al.* (1994). They grafted wild type tobacco plants onto transgenic tobacco expressing the *nahG* gene. Because the *nahG* plants do not accumulate salicylic acid and do not exhibit systemic induced resistance after infection, the authors hypothesized that inoculation of the *nahG* rootstock would not result in the induction of resistance in a wild-type scion if salicylic acid were the translocated signal. They found that the *nahG* rootstock was perfectly capable of generating a signal that resulted in the expression of resistance in the scion. Thus, they also concluded that salicylate was not the translocated signal. Similar experiments with tobacco plants that were suppressed in PAL activity (and also did not accumulate salicylic acid) also supported, albeit indirectly, the same conclusions (Pallas *et al.*, 1996). Recent studies in tobacco (Shulaev *et al.* 1995) and cucumber (Molders *et al.*, 1996) have demonstrated that some of the salicylic acid may be transported out of the infected leaf to the upper parts of the plant. Although these results do not prove that salicylic acid is the only translocated signal, they do show that some of the increased amounts of salicylic acid observed in tissues that are systemic to the infected leaf may be from the site of induction. These results, however, do not rule out the presence of a primary signal that is involved in the systemic induction of salicylic acid synthesis.

4.5. Salicylic Acid Signal Transduction

Understanding the mechanism by which salicylic acid induces resistance is important in determining not only the precise role of this compound in resistance but also how apparently different types of resistance (e.g., parasite specific versus induced) utilize similar defenses. Tobacco contains a soluble protein that binds salicylic acid, and thus could be a "receptor" for this signal (Chen and Klessig, 1991). This protein was determined to be a catalase whose activity was inhibited by salicylic acid (Chen *et al.*, 1993). They hypothesized that salicylic acid induced resistance by inhibition of catalase, which in turn allowed for the localized accumulation of hydrogen peroxide. It was the hydrogen peroxide that then induced resistance. This was supported by the observation that hydrogen peroxide would induce a marker for induced resistance, the pathogenesis-related protein PR1. Further evaluation of this hypothesis by others, however, has failed to support a role for salicylate-mediated hydrogen peroxide accumulation as a mechanism for induced resistance signal transduction (Bi *et al.*, 1995; Neuenschwander *et al.*, 1995). In fact, it has been found that hydrogen peroxide actually induces salicylic acid synthesis (Leon *et al.*, 1995a). However, the observation that hydrogen peroxide stimulates the accumulation of salicylic acid suggests that hydrogen peroxide may be involved, through the induction of salicylic acid accumulation, in the local induction of defense genes in cells surrounding hypersensitive reaction cells (Levine *et al.*, 1994).

In an attempt to determine the role of salicylic acid in resistance induction, a genetic approach has been taken. Several mutants have been identified in Arabidopsis that do not respond to salicylic acid treatment (Weyman *et al.*, 1995; Cao *et al.*, 1994) or constitutively express induced resistance (Bowling *et al.*, 1994; Dietrich *et al.*, 1994; Greenberg *et al.*, 1994). Understanding the function of the mutated genes will be invaluable in understanding the role of salicylic acid in the induction of resistance.

5. APPLICATIONS OF PLANT DEFENSE IN DISEASE CONTROL

The recent cloning of both resistance and defense genes as well as further knowledge of the systemic induced resistance responses of plants has provided several new avenues for disease control. These new approaches use the power of both genetic engineering and the knowledge that all plants have the genes needed for successful defense if these genes can be activated quickly after infection.

5.1. Use of Transgenic Plants

As a result of defense studies, a number of gene products with antipathogen activity have been identified (Linthorst, 1991). Via plant transformation, several of the PR protein genes have been introduced into plants under the control of a constitutive promoter. In some cases the expression of these genes provided a satisfactory level of protection from disease (Broglie *et al.*, 1991). However, in most cases the level of resistance has only been enhanced by a small increment (e.g., Alexander *et al.*, 1993). Although this would seem to be a disappointing result, it should be remembered that the genes being used for transformation represent only a small fraction of the overall defenses employed to combat infection. Nonetheless, these experiments have provided valuable information on the relative contribution of these genes in defense against specific pathogens.

As described earlier, phytoalexins are thought to be important factors in resistance expression, and engineering plants to express phytoalexins may provide another disease control strategy. Transformation of tobacco with the stilbene synthase gene from grapevine resulted in the production of the stilbene phytoalexin resveratrol (Hahn *et al.*, 1993). The transformed tobacco plants expressed a higher degree of resistance to infection by the fungal pathogen *Botrytris cinerea* as compared with controls. Engineering new secondary metabolism to enhance resistance may, however, not come completely without surprises. Zook *et al.* (1996a,b) examined cultured tobacco cells transformed with a sesquiterpene cyclase from the fungus *Fusarium sporotrichioides*. They found that the cells did produce the new sesquiterpene without compromising the ability of the cells to produce the normal sesquiterpene phytoalexins. However, the transformed tobacco cells also produced a completely novel compound that was derived from the fungal sesquiterpene cyclase product. Thus, although engineering new secondary metabolism as an approach to enhance resistance is possible, it also appears that the plant may convert the engineered metabolites into other compounds whose structure cannot be predicted.

Understanding the mechanisms that pathogens use to infect tissue may also lead to new methods of disease control by engineering plants that nullify the pathogenicity factors. An example where this principle may be first utilized is in the control of the white mold disease of canola and other crops caused by *Sclerotinia sclerotiorum*. The simple organic acid oxalic acid is an important pathogenicity factor in the broad-host-range pathogen *S. sclerotiorum* (Godoy *et al.*, 1990). Mutants of this fungus that can no longer produce this acid are no longer pathogenic. Thompson *et al.* (1995) isolated the gene that degrades oxalic acid, oxalate oxidase, from barley roots and used this gene to transform *Brassica napus* (canola), a crop plant that is susceptible to this fungus. Canola expressing this gene were much more resistant to treatment with oxalic acid than were the nontransformants and thus may be more resistant to *Sclerotinia*.

5.2. Applications of Induced Resistance

Another possible strategy of plant disease control that has its roots in a basic understanding of the plant's response to infection is the induced or acquired resistance response. As described earlier, prior infection of plants with pathogens can often result in a localized or systemic increase in resistance of the plant to further infection. Because this resistance can protect the plant against a number of pathogens and uses the plant's natural defenses, it has been suggested that induced resistance should be an environmentally safe as well as effective control measure (Kuć *et al.*, 1995a). A major problem with the use of induced resistance, however, has been in how to implement this type of control in a practical way.

There have been two general approaches to using induced resistance in practical disease control. The first has been to search for chemicals that are not toxic to the pathogen, but have the capacity to induce resistance in the plant to subsequent pathogen infection (Kessmann *et al.*, 1994). The general idea for using chemicals to induce resistance is not new (see Hammerschmidt and Becker, 1997), but it is an attractive approach because growers could apply inducing agents in much the same way they apply traditional pesticides. The second approach is to search for microbes that are not pathogens, but have the ability to interact with plants and induce resistance (Tuzun and Kloepper, 1995).

The use of synthetic chemical activators or inducers of resistance appears to be a practical reality. Benzothiadiazole (BTH, Fig. 4) has been reported to activate resistance in wheat (Gorlach et al., 1996), tobacco (Friedrich et al., 1996), and Arabidopsis (Lawton et al., 1996). The reports indicate that application of BTH results in molecular and biochemical changes that are the same as found in pathogen-induced resistance. 2,6-Dichloroisonicotinic acid (INA; Fig. 4) (Métraux et al., 1991) also induces resistance via stimulation of the same genes involved in pathogen-induced resistance. Resistance to multiple pathogens after treatment of bean (*Phaseolus vulgaris*) has also been reported (Dann and Deverall, 1995). Several years of field trials with both INA and BTH demonstrated that induced resistance can effectively reduce disease in soybean and cucumber (Hammerschmidt, Dann, Diers, and Widders, unpublished results). Similarly, field trials with bean have shown that INA can effectively control rust infection (Dann and Deverall, 1996). Thus, the practical application of induced resistance via chemical induction has been demonstrated.

Another approach is to use nonpathogenic microbes to induce resistance. A group of bacteria known as plant growth-promoting rhizobacteria (PGPR) have been reported to enhance resistance to subsequent pathogen infection (Tuzun and Kloepper, 1995). Similarly, infection of potato by mycorrhizal fungi will also protect the tubers from a rot disease (Niemira et al., 1996). Although the

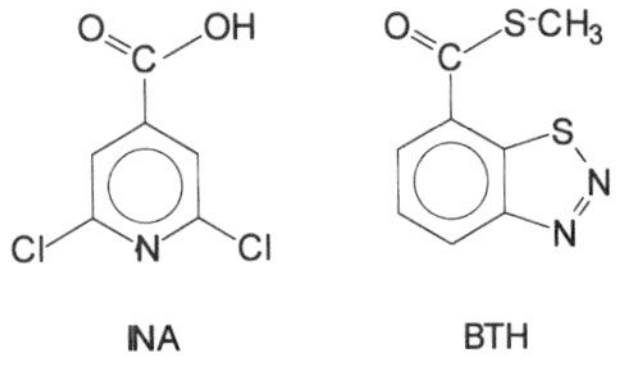

FIGURE 4. Synthetic chemical activators of disease resistance. INA, 2,6-dichloroisonicotinic acid; BTH, benzo[1,2,3]thiadiazole-7-carbothioic acid-S-methyl ester.

actual mechanism of protection induced by these pathogens is not known, the results suggest that some form of induced resistance is part of the effect because of the spatial separation between the location of the inducing organisms and the tissues later challenge inoculation with the pathogen. However, the nature of the induced resistance state, at least in one case, does not appear to be like classical induced resistance. Inoculation of roots of Arabidopsis with *Pseudomonas fluorescens* strain WCS417r resulted in the systemic protection of the foliage against the fungal pathogen *Fusarium oxysporum* f.sp. *raphini* and the bacterial pathogen *Pseudomonas syringae* pv. *tomato* (Pieterse *et al.*, 1996). Contrary to what would be expected in classical induced resistance, inoculation of strain WCS417r onto roots of Arabidopsis expressing the *nahG* gene also expressed induced resistance to the two pathogens. These induced plants also did not exhibit PR gene expression. Thus, the resistance induced by this nonpathogenic bacterium appeared to be mediated by a mechanism that did not involve salicylic acid signal transduction or the expression of PR proteins. Further studies are needed to determine how the plant is able to restrict the development of the pathogen before the nature of this induced resistance can be understood. However, use of another *P. fluorescens* strain (CHAO) on tobacco did induce resistance that was associated with PR gene expression and salicylic acid accumulation (Maurhofer *et al.*, 1994). This study also used a mutant of strain CHAO that did not produce antibiotics, and thus was able to exclude transport of antibiotics produced by the pathogen as the mechanism of the observed resistance. Regardless of the mechanisms, the use of nonpathogenic plant-associated microbes may provide another means of practical disease control that is based on understanding basic mechanisms of plant–microbe interactions.

6. CONCLUSIONS

Plants are the potential targets for infection by a wide array of pathogens. However, recognition systems and a complex battery of defenses have evolved to provide the plant with excellent means of defense against all but a few pathogens. Current research on the use of transgenic plants as well as knowledge of inducible defenses presents us with the opportunity to provide a new generation of plant disease controls that are based on natural resistance mechanisms that have the potential of being environmentally sound and compatible with traditional disease control strategies.

ACKNOWLEDGMENTS. The author's research was supported in part by the Michigan Agricultural Experiments Station, and a grant from the USDA/NRICGP and NSF (IBN-9220912). The author thanks B.A. Dankert for comments on the manuscript.

REFERENCES

Abad, L. R., Durzo, M. P., Liu, D., Narasimhan, M. L., Reuveni, M., Zhu J. K., Niu, X. M., Singh, N. K., Hasegawa, P. M., and Bressan, R. A., 1996, Antifungal activity of tobacco osmotin has specificity and involves plasma-membrane permeabilization, *Plant Sci.* **118**:11–23.

Agrios, G. N., 1988, *Plant Pathology*, 3rd ed., Academic Press, San Diego.

Aist, J. R., 1983, Structural responses as resistance mechanisms. in: *The Dynamics of Host Defence* (J. A. Bailey and B. J. Deverall, eds.), Academic Press, Sydney, pp. 33–70.

Alexander, D., Goodman, R. M., Gut-Rella, M., Glascock, C., Weyman, K., Friedrich, L., Maddox, D., Ahl-Goy, P., Luntz, T., Ward, E., and Ryals, J., 1993, Increased tolerance to two oomycete pathogens in transgenic tobacco expressing pathogenesis-related protein 1a, *Proc. Natl. Acad. Sci. USA* **90**:7327–7331.

Appel, H. M., 1993, Phenolics in ecological interactions: The importance of oxidation, *J. Chem. Ecol.* **19**:1521–1552.

Baker, C. J., and Orlandi, E. W., 1995, Active oxygen in plant pathogenesis, *Annu. Rev. Phytopathol.* **33**:299–321.

Bi, Y.-M., Kenton, P., Mur, L., Darby, R. L., and Draper, K., 1995, Hydrogen peroxide does not function downstream of salicylic acid in the induction of PR protein expression, *Plant J.* **8**:235–245.

Bowling, S. A., Guo, A., Gordon, A. S., Klessig, D. F., and Dong, X., 1994. A mutation in Arabidopsis that leads to constitutive expression of systemic resistance, *Plant Cell* **6**:1845–1857.

Bowyer, P., Clarke, B. R., Lunness, P., Daniel, M. J., and Osbourn, A. E., 1995, Host range of a plant pathogenic fungus determined by a saponin detoxifying enzyme, *Science* **267**:371–374.

Brisson, L. F., Tenhaken, R., and Lamb, C., 1994, Function of oxidative cross-linking of cell wall structural proteins in plant disease resistance, *Plant Cell* **6**:1703–1712.

Broglie, K., Chet, I., Holliday, M., Cressman, R., Riddle, P., Knowlton, S., Mauvias, C. J., and Broglie, R., 1991, Transgenic plants with enhanced resistance to fungal pathogens, *Science* **254**:1194–1197.

Cao, H., Bowling, S. A., Gordon, A. S., and Dong, X., 1994, Characterization of an Arabidopsis mutant that is non-responsive to inducers of systemic acquired resistance, *Plant Cell* **6**:1583–1592.

Chen, Z., and Klessig, D. F., 1991, Identification of a soluble salicylic-acid binding protein that may function in signal transduction in the plant disease resistance response, *Proc. Natl. Acad. Sci. USA* **88**:8179–8183.

Chen, Z., Silva, H., and Klessig, D. F., 1993, Active oxygen species in the induction of plant systemic acquired resistance by salicylic acid, *Science* **262**:1883–1885.

Croft, K. P. C., Juttner, F., and Slusarenko, A. J., 1993, Volatile products of the lipoxygenase pathway evolved from *Phaseolus vulgaris* L. leaves inoculated with *Pseudomonas syringae* pv. *phaseolicola*, *Plant Physiol.* **101**:13–24.

Cutt, J. R., and Klessig, D. F., 1992, Pathogenesis-related proteins, in: *Plant Gene Research* (F. Meins and T. Boller, eds.), Springer-Verlag, Berlin, pp. 181–216.

Dann, E. K., and Deverall, B. J., 1995, Effectiveness of systemic resistance in bean against foliar and soil-borne pathogens as induced by biological and chemical means, *Plant Pathol.* **44**:458–466.

Dann, E. K., and Deverall, B. J., 1996, 2,6-Dichloroisonicotinic acid (INA) induced resistance in green beans to the rust pathogen, *Uromyces appendiculatus*, under field conditions, *Aust. Plant Pathol.* **25**:199–204.

Dean, R. A., and Kuć, J., 1986a, Induced systemic protection in cucumber: The source of the "signal," *Physiol. Mol. Plant Pathol.* **28**:227–233.

Dean, R. A., and Kuć, J., 1986b, Induced systemic protection in cucumber: Time of production and movement of the signal, *Phytopathology* **76**:966–970.

Delaney, T. P., Uknes, S., Vernooij, B., Friedrich, L., Weymann, K., Negrotto, D., Gaffney, T., Gut-Rella, M., Kessmann, H., Ward, E., and Ryals, J., 1994, A central role of salicylic acid in plant disease resistance, *Science* **266**:1247–1250.

Dietrich, R. A., Delaney, T.P., Uknes, S. J., Ward, E. R., Ryals, J. A., and Dangl, J. L., 1994, Arabidopsis mutants simulating disease resistance response, *Cell* **77**:565–577.

Dixon, R. A., Harrison, M. J., and Lamb, C. J., 1994, Early events in the activation of plant defense responses, *Annu. Rev. Phytopathol.* **32**:479–501.

Enkerli, J., Gisi, U., and Mosinger, E., 1993, Systemic acquired resistance to *Phytophthora infestans* in tomato and the role of pathogenesis related proteins, *Physiol. Mol. Plant Pathol.* **43**:161–171.

Essenberg, M., Pierce, M. L., Cover, E. C., Hamilton, B., Richardson, P. E., and Scholes, V. E., 1992, A method for determining phytoalexin concentrations in fluorescent hypersensitively necrotic cells in cotton leaves, *Physiol. Mol. Plant Pathol.* **41**:101–109.

Farmer, E. E., 1994, Fatty acid signalling in plants and their associated microorganisms, *Plant Mol. Biol.* **26**:1423–1437.

Friedrich, L., Lawton, K., Ruess, W., Masner, P., Specker, N., Rella, M. G., Meier, B., Dincher, S., Staub, T., Uknes, S., Metraux, J. P., Kessmann, H., and Ryals, J., 1996, A benzothiadiazole derivative induces systemic acquired resistance in tobacco, *Plant J.* **10**:61–70.

Gaffney, T., Friedrich, L., Vernooij, B., Negrotto, D., Nye, G., Uknes, S., Ward, E., Kessmann, H., and Ryals, J., 1993, Requirement of salicylic acid for the induction of systemic acquired resistance, *Science* **261**:754–766.

Glazebrook, J., and Ausubel, F. M., 1994, Isolation of phytoalexin deficient mutants of *Arabidopsis thaliana* and characterization of their interactions with bacterial pathogens, *Proc. Natl. Acad. Sci. USA* **91**:8955–8959.

Godoy, G., Steadman, J. R., Dickman, M. B., and Dam, R., 1990, Use of mutants to demonstrate the role of oxalic acid in pathogenicity of *Sclerotinia sclerotiorum* on *Phaseolus vulgaris*, *Physiol. Mol. Plant Pathol.* **37**:179–191.

Goodman, R. N., and Novacki, A. J., 1994, *The Hypersensitive Reaction in Plants to Pathogens*, APS Press, St. Paul, p. 244. Gorlach, J., Volrath, S., Knauf-Beiter, G., Hengy, G., Beckhove, U., Kogel, K. H., Oostendorp, M., Staub, T., Ward, E., Kessmann, H., and Ryals, J., 1996, Benzothiadiazole, a novel class of inducers of systemic acquired resistance, activates gene expression and disease resistance in wheat, *Plant Cell* **8**:629–643.

Greenberg, J. T., Guo, A., Klessig, D. F., and Ausubel, F. M., 1994, Programmed cell death in plants: A pathogen triggered response activated coordinately with multiple defense functions, *Cell* **77**:551–563.

Hahn, R., Reif, H. J., Krause, E., Langebartels, R., Kindl, H., Vornam, B., Wiese, W., Schmelzer, E., Schreier, P. H., Stocker, R. H., and Stenzel, K., 1993, Disease resistance results from foreign phytoalexin expression in a novel plant, *Nature* **361**:153–156.

Hammerschmidt, R., 1993, The nature and generation of systemic signals induced by pathogens, arthropod herbivores and wounds, *Adv. Plant Pathol.* **10**:307–337.

Hammerschmidt, R., and Becker, J. S., 1997, Acquired resistance to disease. *Hortic. Rev.* **18**:247–289.

Hammerschmidt, R., and Kuć, J., 1982, Lignification as a mechanism for induced systemic resistance in cucumber, *Physiol. Plant Pathol.* **20**:61–71.

Hammerschmidt, R., and Kuć, J. (eds.), 1995, *Induced Resistance to Disease in Plants*, Kluwer, Dordrecht.

Hammerschmidt, R. and Schultz, J. C., 1996, Multiple defenses and signals in plant defense against pathogens and herbivores, *Recent Adv. Phytochem.* **30**:121–154.

Hammerschmidt, R., Nuckles, E. M., and Kuć, J., 1982, Association of enhanced peroxidase activity with induced systemic resistance of cucumber to *Colletotrichum lagenarium*, *Physiol. Plant Pathol.* **20**:73–82.

Hammerschmidt, R., Bonnen, A. M., Baker, K. K., and Bersgstom, G. C., 1985, Association of epidermal lignification with nonhost resistance of cucurbits to fungi, *Can. J. Bot.* **63**:2393–2398.

Heath, M. C., 1994, Evolution of resistance to fungal parasitism in natural ecosystems, *New Phytol.* **119**:331–343.

Heath, M. C., 1995, Thoughts on the role and evolution of induced resistance in natural ecosystems, and its relationship to other types of plant defenses against disease, in: *Induced Resistance to Disease in Plants* (R. Hammerschmidt and J. Kuć, eds.), Kluwer, Dordrecht, pp. 141–151.

Kessmann, H., Staub, T., Hofmann, C., Maetzke, T., Herzog, J., Ward, E., Uknes, S., and Ryals, J., 1994, Induction of systemic acquired disease resistance in plants by chemicals, *Annu. Rev. Phytopathol.* **32**:439–459.

Klessig, D. A., and Malamy, J., 1994, The salicylic acid signal in plants, *Plant Mol. Biol.* **26**:1439–1458.

Kuć, J., 1995a, Induced systemic resistance—an overview, in: *Induced Resistance to Disease in Plants* (R. Hammerschmidt and J. Kuć, eds.), Kluwer, Dordrecht, pp. 169–175.

Kuć, J., 1995b, Phytoalexins, stress metabolism and disease resistance in plants, *Annu. Rev. Phytopathol.* **33**:275–297.

Lamb, C., 1996, A ligand–receptor mechanism in plant–pathogen recognition, *Science* **274**:2038–2039.

Lawton, K. A., Friedrich, L., Hunt, M., Weymann, K., Delaney, T., Kessmann, H., Staub, T., and Ryals, J., 1996, Benzothiadiazole induces disease resistance in Arabidopsis by activation of the systemic acquired-resistance signal transduction pathway. *Plant Journal* **10**:71–82.

Lazarovits, G., and Ward, E.W.B., 1982, Polyphenoloxidase activity in soybean hypocotyls at sites inoculated with *Phytophthora megasperma* f.sp. *glycinea*, *Physiol. Plant Pathol.* **21**:227–236.

Leach, J. E., and White, F. F., 1996, Bacterial avirulence genes, *Annu. Rev. Phytopathol.* **34**:153–179.

Leon, J., Yalpani, N., Raskin, I., and Lawton, M. A., 1993, Induction of benzoic acid 2-hydroxylase in virus-inoculated tobacco, *Plant Physiol.* **103**:323–328.

Leon, J., Lawton, M. A., and Raskin, I., 1995a, Hydrogen-peroxide stimulates salicylic acid biosynthesis in tobacco, *Plant Physiol.* **108**:1673–1678.

Leon, J., Shulaev, V., Yalpani, N., Lawton, M. A., and Raskin, I., 1995b, Benzoic acid 2-hydroxylase, a soluble oxygenase from tobacco, catalyzes salicylic acid biosynthesis, *Proc. Natl. Acad. Sci. USA* **92**:10413–10417.

Levine, A., Tenhaken, R., Dixon, R. A., and Lamb, C., 1994, H_2O_2 from the oxidative burst orchestrates the plant hypersensitive disease resistance response, *Cell* **79**:583–593.

Levine, A., Pennell, R. I., Alverez, M. E., Palmer, R., and Lamb, C., 1996, Calcium-mediated apoptosis in a plant hypersensitive disease resistance response, *Curr. Biol.* **6**:427–437.

Linthorst, H. J. M., 1991, Pathogenesis-related proteins in plants, *Crit. Rev. Plant Sci.* **10**:123–150.

Liu, D., Raghothama, K. G., Hasegawa, P. M., and Bressan, R. A., 1994, Osmotin overexpression in potato delays development of disease symptoms, *Proc. Natl. Acad. Sci. USA* **91**:1888–1892.

Maher, E. A., Bate, N. J., Ni, W., Elkind, Y., Dixon, R. A., and Lamb, C. J., 1994, Increased disease susceptibility of transgenic tobacco plants with suppressed levels of preformed phenylpropanoid products, *Proc. Natl. Acad. Sci. USA* **91**: 7802–7806.

Malamy, J., Carr, J. P., Klessig, D. F., and Raskin, I., 1990, Salicylic acid—A likely endogenous signal in the resistance response of tobacco to tobacco mosaic virus, *Science* **250**:1002–1004.

Mauch-Mani, B., and Slusarenko, A., 1996, Systemic acquired resistance in *Arabidopsis thaliana* induced by a predisposing infection with a pathogenic isolate of *Fusarium oxysporum*, *Mol. Plant–Microbe Interact.* **7**:378–383.

Maurhofer, M., Hase, C., Meuwly, P., Métraux, J. P., and Defago, G., 1994, Induction of systemic resistance of tobacco to tobacco necrosis virus by root-colonizing *Pseudomonas fluorescens* strain CHAO: Influence of the gacA gene and of pyoverdine production, *Phytopathology* **84**:139–146.

Mayama, S., and Tani, T., 1982, Microspectrophotometric analysis of the location of avenalumin accumulation in response to fungal infection, *Physiol. Plant Pathol.* **21**:141–149.

Mayama, S., Bordin, A. P. A., Morikawa, T., Tampo, H., and Kato, H. 1995. Association of avenalumin accumulation with co-segregation of victorin sensitivity and crown rust resistance in oat lines carrying the Pc-2 gene, *Physiol. Mol. Plant Pathol.* **46**:263–274.

Meeley, R. B., Johal, G. S., Briggs, S. P., and Walton, J. D., 1992, A biochemical phenotype for a disease resistance gene of maize. *Plant Cell* **4**:71–77.

Mehdy, M. C., 1994, Active oxygen species in plant defense against pathogens, *Plant Physiol.* **105**:467–472.

Métraux, J. P., Signer, H., Ryals, J., Ward, E., Wyss-Benz, M., Gaudin, J., Raschdorf, K., Schmid, E., Blum, W., and Inverardi, B., 1990, Increase in salicylic acid at the onset of systemic acquired resistance in cucumber, *Science* **250**:1004–1006.

Métraux, J. P., Ahl-Goy, P., Staub, T., Speich, J., Steinemann, A., Ryals, J., and Ward, E., 1991, Induced systemic resistance in cucumber in response to 2,6-dichloroisonicotinic acid and pathogens, in: *Advances in the Molecular Genetics of Plant Microbe Interactions*, (H. Hennecke and D. P. S. Verma, eds.), Kluwer, Dordrecht, Vol. 1, pp. 432–439.

Meuwly, P., Molders, W., Buchala, A., and Métraux, J. P., 1995, Local and systemic biosynthesis of salicylic acid in infected cucumber plants, *Plant Physiol.* **109**:1107–1114.

Michelmore, R. W., 1995, Molecular approaches to manipulation of disease resistance, *Annu. Rev. Phytopathol.* **33**:393–427.

Mills, P. R., and Wood, R. K. S., 1984, The effects of polyacrylic acid, acetyl salicylic acid and salicylic acid on resistance of cucumber to *Colletotrichum lagenarium*, *Phytopathol. Z.* **111**:209–216.

Mittler, R., and Lam, E., 1995, *In situ* detection of nDNA fragmentation during a differentiation of tracheary elements in higher plants, *Plant Physiol.* **108**: 489–493.

Moersbacher, B. M., Noll, U., Gorrichon, L., and Reisner, H. J., 1990, Specific inhibition of lignification breaks hypersensitive resistance of wheat to stem rust, *Plant Physiol.* **93**:465–470.

Mölders, W., Buchala, T., and Métraux, J. P., 1996, Transport of salicylic acid in tobacco necrosis virus-infected cucumber plants, *Plant Physiol.* **112**:787–792.

Neuenschwander, U., Vernooij, B., Friedrich, L., Uknes, S., Kessmann, H., and Ryals, J., 1995, Is hydrogen peroxide a second messenger of salicylic acid in systemic acquired resistance? *Plant J.* **8**:227–233.

Nicholson, R. L., and Hammerschmidt, R., 1992, Phenolic compounds and their role in disease resistance, *Annu. Rev. Phytopathol.* **30**:369–389.

Niemira, B. A., Hammerschmidt, R., and Safir, G. R., 1996, Postharvest suppression of potato dry rot (*Fusarium sambucinum*) in prenuclear minitubers by arbuscular mycorrhizal fungal inoculum, *Am. Potato J.* **73**:509–515.

Osbourn, A. E., Clarke, B. R., Lunnes, P., Scott, P. R., and Daniels, M. J., 1994, An oat species lacking avenacin is susceptible to infection by *Gaumanomyces graminis* var. *tritici*, *Physiol. Mol. Plant Pathol.* **45**:457–467.

Pallas, J. A., Paiva, N. L., Lamb, C. J., and Dixon, R. A., 1996, Tobacco plants epigenetically suppressed in phenylalanine ammonia lyase expression do not develop systemic acquired resistance in response to infection by tobacco mosaic virus, *Plant J.* **10**:281–293.

Paxton, J. D., 1981, Phytoalexins—A working redefinition, *Phytopathol. Z.* **101**: 106–109.

Peng, M., and Kuć, J., 1992, Peroxidase-generated hydrogen peroxide as a source of antifungal activity in vitro and on tobacco leaf disks, *Phytopathology* **82**: 696–699.

Pierce, M. L., Coover, E. C., Richardson, P. E., Scholes, V. E., and Essenberg, M.,

1996, Adequacy of cellular phytoalexin concentrations in hypersensitively responding cotton leaves, *Physiol. Mol. Plant Pathol.* **48**:305–324.

Pierpont, W. S., 1994, Salicylic acid and its derivatives in plants: Medicines, metabolism and messenger molecules, *Adv. Bot. Res.* **20**:164–235.

Pieterse, C. M. J., Vanwees, S. C. M., Hoffland, E., Vanpelt, J. A., and Van Loon, L. C., 1996, Systemic resistance in Arabidopsis induced by biocontrol bacteria is independent of salicylic acid accumulation and pathogenesis-related protein gene expression, *Plant Cell* **8**:1225–1237.

Rasmussen, J. B., Hammerschmidt, R., and Zook, M. N., 1991, Systemic induction of salicylic acid accumulation in cucumber after inoculation with *Pseudomonas syringae* pv. *syringae*, *Plant Physiol.* **97**:1342–1347.

Rasmussen, J. B., Smith, J. A., Williams, S., Burkhardt, W., Ward, E., Somerville, S., Ryals, J., and Hammerschmidt, R., 1995, cDNA cloning and systemic expression of an acidic peroxidase associated with systemic acquired resistance in cucumber, *Physiol. Mol. Plant Pathol.* **46**:389–400.

Ride, J. P., 1978, The role of cell wall alterations in resistance to fungi, *Ann. Appl. Biol.* **89**:302–306.

Ryals, J., Uknes, S., and Ward, E., 1994, Systemic acquired resistance, *Plant Physiol.* **104**:1109–1112.

Ryerson, D. E., and Heath, M. C., 1996, Cleavage of nuclear DNA into oligonucleosomal fragments during cell death induced by fungal infection or by abiotic treatments, *Plant Cell* **8**:393–402.

Schäfer, W., 1994, Molecular mechanisms of fungal pathogenicity in plants, *Annu. Rev. Phytopathol.* **32**:461–477.

Scofield, S. R., Tobias, C. M., Rathjen, J. P., Chang, J. H., Lavelle, D. T., Michelmore, R. W., and Staskawicz, B. J., 1996, Molecular basis of gene-for-gene specificity in bacterial speck disease of tomato, *Science* **274**:2063–2065.

Shulaev, V., Leon, J., and Raskin, I., 1995, Is salicylic acid a translocated signal of systemic acquired resistance in tobacco? *Plant Cell* **7**:1691–1701.

Smith, J. A., and Hammerschmidt, R., 1988, Comparative study of acidic peroxidases associated with induced resistance in cucumber, muskmelon and watermelon, *Physiol. Mol. Plant Pathol.* **33**:255–261.

Smith, J. A., Fulbright, D. W., and Hammerschmidt, R., 1991, Rapid induction of systemic resistance in cucumber by *Pseudomonas syringae* pv. *syringae*, *Physiol. Mol. Plant Pathol.* **38**:223–235.

Smith, M. J., Mazzola, E. P., Sims, J. J., Midland, S. L., Keen, N. T., Burton, V., and Stayton, M. M., 1993, The syringolides: Bacterial c-glycosyl lipids that trigger plant disease resistance, *Tetrahedron Lett.* **34**:223–226.

Snyder, B. A., and Nicholson, R. L. 1990. Synthesis of phytoalexins in sorghum as a site specific response to fungal ingress, *Science* **248**:1637–1639.

Snyder, B. A., Leite, B., Hipskind, J., Butler, L. G., and Nicholson, R. L., 1991, Accumulation of sorghum phytoalexins induced by *Colletotrichum graminicola* at the infection site, *Physiol. Mol. Plant Pathol.* **39**:463–470.

Staskawicz, B. J., Ausubel, F. M., Baker, B. J., Ellis, J. G., and Jones, J. D. G., 1995, Molecular genetics of plant disease resistance, *Science* **268**:661–667.

Stein, B. D., Klomparens, K., and Hammerschmidt, R., 1993, Histochemistry and ultrastructure of the induced resistance response of cucumber plants to *Colletotrichum lagenarium*, *J. Phytopathol.* **137**:177–188.

Stermer, B. A., 1995, Molecular regulation of induced systemic resistance, in: *Induced Resistance to Disease in Plants* (R. Hammerschmidt and J. Kuć, eds.), Kluwer, Dordrecht, pp. 111–140.

Sutherland, M. W., 1991, The generation of oxygen radicals during host responses to infection, *Physiol. Mol. Plant Pathol.* **39**:79–94.

Tang, X., Frederick, R. D., Zhou, J., Halterman, D. A., Jia, Y., and Martin, G. B., 1996, Initiation of plant disease resistance by physical interaction of AvrPto and Pto kinase, *Science* **274**:2060–2063.

Thompson, C., Dunwell, J. M., Johnstone, C. E., Lay, V., Ray, J., Schmitt, M., Watson, H., and Nisbet, G., 1995, Degradation of oxalic acid by transgenic oilseed rape plants expressing oxalate oxidase, *Euphytica* **85**:169–172.

Tsuji, J., Jackson, E. P., Gage, D. A., Hammerschmidt, R., and Somerville, S. C., 1992, Phytoalexin accumulation in *Arabidopsis thaliana* during the hypersensitive response to *Pseudomonas syringae* pv. *syringae*, *Plant Physiol.* **98**:1304–1309.

Tuzun, S., and Kloepper, J., 1995, Practical application and implementation of induced resistance, in: *Induced Resistance to Disease in Plants* (R. Hammerschmidt and J. Kuć, eds.), Kluwer, Dordrecht, pp. 152–168.

Van Kan, L. A. L., Vanden Ackeveken, G. F. J. M., and de Wit, P. J. M., 1991, Cloning and characterization of cDNA of avirulence gene avr9 of the fungal pathogen *Cladosporium fulvum*, caused agent of tomato leaf mold, *Mol. Plant–Microbe Interact.* **4**:52–59.

van Loon, L. C., Pierpoint, W. S., Boller, T., and Conejero, V., 1994, Recommendations for naming plant pathogenesis-related proteins, *Plant Mol. Biol. Rep.* **12**:245–264.

Vernooij, B., Friedrich, L., Morse, A., Reist, R., Kolditz-Jawhar, R., Ward, E., Uknes, S., Kessmann, H., and Ryals, J., 1994, Salicylic acid is not the translocated signal responsible for inducing systemic acquired resistance but is required in signal transduction, *Plant Cell* **6**:959–965.

Wang, H., Li, J., Bostock, R. M., and Gilchrist, D. G., 1996, Apoptosis: A functional paradigm for programmed cell death induced by a host-selective phytotoxin and invoked during development, *Plant Cell* **8**:375–391.

Ward, E., Uknes, S. J., Williams, S. C., Dincher, S. S., Wiederhold, D. L., Alexander, D. C., Ahl-Goy, P., Métraux, J. P., and Ryals, J. A., 1991, Coordinate gene activity in response to agents that induce systemic acquired resistance, *Plant Cell* **3**:1085–1094.

Weyman, K., Hunt, M., Uknes, S., Neuenschwender, U., Lawton, K., Steiner, H. Y., and Ryals, J., 1995, Suppression and restoration of lesion formation in Arabidopsis lsd mutants, *Plant Cell* **7**:2013–2022.

White, R. F., 1979, Acetylsalicylic acid (aspirin) induces resistance to tobacco maosaic virus in tobacco, *Virology* **99**:410–412.

Yalpani, N., Leon, J., Lawton, M. A., and Raskin, I., 1993, Pathway of salicylic acid biosynthesis in healthy and virus-inoculated tobacco, *Plant Physiol.* **103**:315–321.

Zeyen, R. J., Bushnell, W. R., Carver, T. L. W., Robbins, M. P., Clark, T. A., Boyles, D. A., and Vance, C. P., 1995, Inhibiting phenylalanine ammonia lyase and cinnamyl-alcohol dehydrogenase suppresses Mla1 (HR) but not mlo5 (non-HR) barley powdery mildew resistances. *Physiol. Mol. Plant Pathol.* **47**:119–140.

Zook, M., and Hammerschmidt, R., 1997, Origin of the thiazole ring in camalexin, a phytoalexin from *Arabidopsis thaliana*, *Plant Physiol.* **113**:463–468.

Zook, M., Hohn, T., Bonnen, A., Tsuji, J., and Hammerschmidt, R., 1996a, Characterization of novel sesquiterpenoid biosynthesis in tobacco expressing a fungal sesquiterpene synthase, *Plant Physiol.* **112:**311–318.

Zook, M., Johnson, K., Hohn, T., and Hammerschmidt, R., 1996b, Structural characterization of 15-hydroxytrichodiene, a sesquiterpenoid produced by transformed tobacco cell suspension cultures expressing a trichodiene synthase gene from *Fusarium sporotrichioides, Phytochemistry* **43:**1235–1237.

3

Stress Responses in *Drosophila* Cells

EIKO AKABOSHI and YUTAKA INOUE

1. INTRODUCTION

The environmental stresses (e.g., pollution, radiation, and chemicals) increase in the modern society. Various kinds of stress attack all living organisms. All organisms appear to have innate ability to overcome these stresses (sometimes threats) for survival. Their sensitivity and responses, however, are varied, depending on the organism and on the developmental stage, age, tissue, and cell. Immune and inflammation reactions also play important roles in the processes in which organisms recover from the stress and rescue themselves (for reviews, see Cociancich *et al.*, 1994; Hultmark, 1993; Wilder, 1995). When organisms are exposed to above the threshold threats, they either die or survive with severe damage. The survivors, lucky as it may sound, must undergo sustained agonies. Further, the damage in their DNA, either in eggs or spermatozoa, may result in the appearance of mutated or aberrant progeny. An increasingly large number of people recognize such stresses as a serious problem for the society and for all

EIKO AKABOSHI • Institute for Molecular and Cellular Biology, Osaka University, 1-3 Yamadaoka, Suita 565, Japan.   YUTAKA INOUE • Osaka University of Foreign Studies, Minoo, Osaka 562, Japan.

Stress-Inducible Processes in Higher Eukaryotic Cells, edited by Koval. Plenum Press, New York, 1997.

living organisms. Thus, a worldwide campaign to keep the earth clean and safe has been boosted. Accordingly, the stress and stress-response have emerged recently as an important topic in cell biology.

The fruit fly, *Drosophila*, which played an indispensable role in establishing classical genetics, is still being extensively used as a model system in modern biochemical and molecular studies. It played a leading role in heat shock (HS) research and subsequently in general stress research. Puffs are observed in the salivary gland polytene chromosomes of the third-instar larvae. The puffs, representing the site (locus) of active RNA transcription, can be artificially induced to change. A unique set of puffs on polytene chromosomes of *D. busckii* were induced by HS, nitrophenol, and sodium salicylate (Ritossa, 1962). Such altered puffing activities were also observed to occur with various other agents (review: Ashburner and Bonner, 1979). These works pioneered the development of the so-called HS or stress response studies, one of the major recent topics in cell biology, biochemistry, and molecular biology. Studies on stress-inducible genes and gene products had been done solely with *Drosophila*, until the end of the 1970s when a HS response was reported in the chicken embryo fibroblast (Kelley and Schlesinger, 1978), and in *Escherichia coli* and very soon afterwards in various other organisms including mammals and plants (review: Lindquist, 1986). The HS response is now known to be a universal property of all living organisms. The temperature an organism recognizes as HS varies, depending on the optimum growing temperature of the organism. Genes coding for the major heat shock proteins (HSPs) were cloned and sequenced in *Drosophila* in the late 1970s and early 1980s (review: Nover, 1991). The identification, isolation, and sequence analysis of the HS genes have enabled the study of the expression of HS genes and their regulatory mechanisms. Following successful cloning of the HS cDNAs and genomic DNAs in *Drosophila*, the cloning and sequence analysis of major HS genes in other organisms were also carried out. The HSPs have been confirmed to be among the most highly conserved proteins in nature. Furthermore, most HSPs are found to be inducible by other stresses, and thus HSPs are sometimes called stress proteins.

When *Drosophila* cultured cells are exposed to stress caused by various agents, the cells respond with a variety of changes such as synthesis, degradation, modulation, and distribution of macro-molecules (e.g., DNA, RNA, protein, lipid) and often concentrations of other substances (metals, ions and small molecules) are also affected. In this chapter, stress-inducible processes in *Drosophila* cultured cells will be examined paying particular attention to the mRNA induced by various DNA-damaging agents, especially methyl methanesulfonate.

When eukaryotic cells are exposed to agents that cause alterations of DNA, many genes and/or proteins are activated or induced. DNA damage, if not repaired, may lead to ailments, cancer, and genetic disorders. Organisms have pathways by which they protect themselves and their offspring from such threats. In *Drosophila* cultured cells treated with a variety of DNA-damaging agents, many new protein species are induced to express, depending on the agent, along with the so-called HSPs or stress proteins (Akaboshi and Howard-Flanders, 1989). These results suggest that various sets of genes respond differently to various stresses. Inducible DNA repair pathways as recognized in bacteria and in lower eukaryotes such as yeast, however, have not been confirmed in *Drosophila*. Rather, the mode of induction in *Drosophila* appears to be similar to that in mammalian cells. We will discuss how higher eukaryotes overcome the stress caused by DNA-damaging agents, by comparing the results obtained in *Drosophila* cells with those in mammalian cells.

2. EXPERIMENTAL SYSTEMS

2.1. Culture of *Drosophila* Cells and Application of Stress Agents

A subline SICII of *Drosophila*, which was selected from the cell line GM1 for small cell size, was used in our system. The cells were grown in monolayers at 25°C in M3 (Bf) medium supplemented with 10% heat-inactivated (at 56°C for 30 min) fetal bovine serum, penicillin G (200 U/ml), and streptomycin (50 μg/ml), and without an external supply of CO_2 gas.

The cells were seeded into 90-mm tissue culture plates at densities of $1-2 \times 10^7$ cells/plate. After a 36-hr incubation, the cells were treated with chemicals or incubated at 37°C for HS treatment. Ultraviolet (UV) irradiation was carried out after the medium was removed and replaced with *Drosophila* Ringer solution (Akaboshi and Howard-Flanders, 1989).

2.2. Analysis of Induced mRNA

2.2.1. RNA Preparation, in Vitro Translation, and 2D Electrophoresis

Total RNA was extracted from the treated and untreated cells by the guanidium thiocyanate method. Poly(A)$^+$ RNA was purified by chromatography on oligo(dT)-cellulose [oligo(dT)-cellulose Type 7, Pharmacia].

The poly(A)$^+$ RNA (mRNA fraction) was translated *in vitro* in a reticulocyte lysate system according to the protocols given by the supplier (Amersham). The poly(A)$^+$ RNA (1.8 μg) was incubated in a 50-μl lysate mixture containing [^{35}S]methionine, at 30°C for 60 min. The translation products were analyzed by SDS-polyacrylamide gel electrophoresis (PAGE), or two-dimensional (2D) electrophoresis using isoelectric focusing (IEF) in the first dimension and 12% SDS-PAGE in the second. 2-D Pharmalyte (Pharmacia) was used for IEF according to the protocols given by the supplier (Akaboshi and Howard-Flanders, 1989).

cDNA libraries were constructed by using a cDNA synthesis system (Amersham), and ligating to Lambda ZAP vector DNA (Stratagene, USA).

2.2.2. Enrichment of Induced mRNA

cDNAs corresponding to the mRNAs induced by the treatment with stress reagents were enriched by the subtraction method (Welcher *et al.*, 1986). cDNAs labeled with ^{32}P-CTP were synthesized from mRNA fractions from the treated and untreated cells separately. At the same time, the noninduced mRNA fraction was labeled with photobiotin (Vector Labs). The biotinylated RNA and ^{32}P-labeled treated cDNA were hybridized in 0.65 M NaCl, 0.04 M NaPO$_4$ pH 6.8, 1 mM EDTA, and 0.05% SDS at 60°C for 20 hr. The samples were incubated with Avidin-DN (Vector Labs) and chromatographed on a copper-chelate resin that was adjusted from aminoacetic acid–agarose (Pierce Chemical) by passing cupric sulfate solution through the column. The procedure of the solution hybridization–copper-chelate chromatography was repeated to obtain enriched probe. The treated cDNA enriched for stress-induced mRNA thus obtained was then used in plaque hybridization with cDNA from untreated mRNA as probes. The same ^{32}P-labeled probes can be used repeatedly.

2.2.3. Mapping of Induced cDNA on Chromosomes

The cDNAs were first labeled with biotin-14-dATP by nick translation. The biotinylated DNA probes were hybridized *in situ* to the *Drosophila* larval salivary gland polytene chromosomes (Engels *et al.*, 1986).

3. COMPLEX RESPONSE TO VARIOUS AGENTS IN *DROSOPHILA* CELLS

3.1. Effect of Various Agents on Cells

The *Drosophila* cells were treated with various DNA-damaging agents [methyl methanesulfonate (MMS), ethyl methanesulfonate (EMS), 4-nitroquinoline N-oxide (4-NQO), mitomycin C (MMC), hydrogen peroxide, angelicin plus near-UV (UVA) irradiation, and germicidal UV (UVC) irradiation], a DNA gyrase inhibitor (nalidixic acid), a tumor promoter, 12-O-tetradecanoyl-phorbol-13-acetate (TPA), and HS. mRNA fractions prepared from treated and untreated cells were translated *in vitro* and the translation products were analyzed by SDS-PAGE. The results showed that each agent induced or enhanced specific bands as well as common bands. MMS, angelicin, and HS were potent inducers. EMS, TPA, and 4-NQO formed a middle class of induction activity. Nalidixic acid, MMC, and hydrogen peroxide were poor inducers (data not shown). SDS-PAGE can thus be successfully used to detect the presence of inducible mRNAs through newly appearing proteins.

Next, the translation products were analyzed by 2D electrophoresis. MMS was the most potent inducer in terms of numbers and intensity of protein spots among the agents examined, as expected from SDS-PAGE. The results are shown in Fig. 1. In comparison with the untreated mRNA fraction (A), at least 25 new or larger spots appeared, and many others were reduced (B).

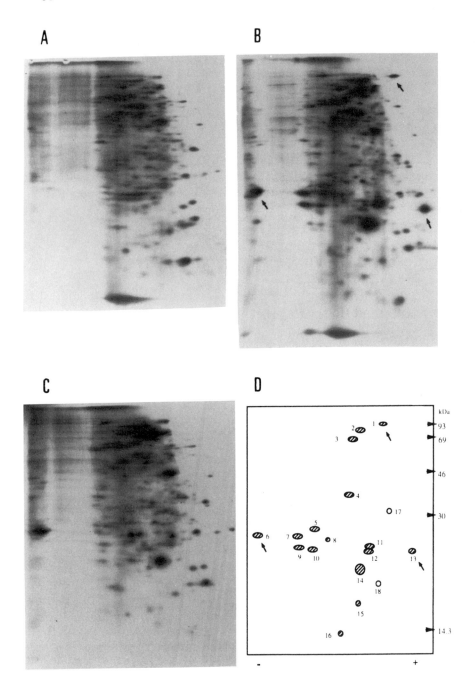

Several spots were coincident with the ones induced by HS (C) (Akaboshi and Howard-Flanders, 1989).

Table I summarizes the appearance of the main spots induced by various agents. EMS is a poor inducer compared with MMS, although both are alkylating agents and EMS is a more effective mutagen for *D. melanogaster*. This observation might be related to the fact that in *Drosophila*, EMS causes production of point mutation(s) more efficiently than MMS, whereas MMS is effective in producing deletion mutations. Angelicin induces most of the protein spots induced by MMS treatment, but to a lesser degree. 4-NQO is a good inducer, and at the same time suppresses the normal spots most severely among the agents examined (Akaboshi, unpublished). MMC, UV, and hydrogen peroxide were poor inducers and their weak inducibility was confirmed under various conditions (concentration, dose, and time). Protein spots 2, 3, 6, 8, and 14, induced by various agents, are the same as HSPs. Spots 3, 6, and 14 probably represent the 70k family, the 26k, and the 23k HSP, respectively, from a comparison with the pattern of HSPs (Akaboshi and Howard-Flanders, 1989).

3.2. Heat Shock or Stress Proteins

3.2.1. Expression and Function of Major HSPs

Regulation and expression of HS genes and distribution and function of HSPs have been extensively studied and earlier literature has been reviewed by Lindquist (1986) and by Ashburner and Bonner (1979). Genomic organization for major HSPs of *Drosophila* has been reviewed by Nover (1991). Here, activation of HS genes and functions of their products (HSPs) will be discussed briefly.

←————————————————————————————————

FIGURE 1. Two-dimensional electrophoresis of *in vitro* translation products of poly(A)$^+$ RNA from *Drosophila* cultured cells treated with MMS. Cells were treated without (A) or with 0.02% MMS for 6 hr (B). After changing medium, the cells were incubated for a further 7 hr. Total RNA was extracted, and poly(A)$^+$ RNA was purified and translated *in vitro* in a reticulocyte lysate system as described in Section 2.2.1. Translation products were analyzed by two-dimensional electrophoresis. The arrows show some examples of new or enhanced spots. (C) Cells were exposed to 37°C for 2 hr. (D) Induced spots are illustrated and numbered. Striped spots and plain ones show proteins induced by MMS and specific proteins induced significantly by other agents, respectively. (From Akaboshi and Howard-Flanders, 1989, with permission.)

TABLE I
Summary of Proteins Induced by Various Agents[a,b]

	1	2	3	4	5	6	7	8	9	10	11	12	13	14	15	16	17	18
(a) MMS	+	+	+	+	+	+	+	+	+	+	+	+	+	+	+	+		
(b) EMS		+		+		+		+			+	+	+	+				
(c) AG	+	+	+		+	+	+	+		+	+	+	+	+				
(d) MMC		+		+				+	+		+	+		+				
(e) 4-NQO		+	+		+	+	+	+	+		+	+	+	+				+
(f) UV		+						+				+	+	+				
(g) HP			+			+		+			+	+		+				
(h) NA			+					+				+				+		
(i) TPA		+					+		+			+		+				
(j) HS		+	+			+		+							+			

[a]Data from Akaboshi and Howard-Flanders (1989) with permission.
[b](a) *Drosophila* cells were treated with 0.02% MMS as described in Fig. 1. (b, d, g) Cells were treated with 0.1% EMS (b), 120 μg/ml MMC (d), or 50 mM hydrogen peroxide (g) for 6–7 hr. (e, h, i) Cells were treated with 5 μg/ml 4-NQO (e), 100 μg/ml nalidixic acid (h), or 0.5 μg/ml TPA (i) for 6 hr, the medium changed and incubated for 14 hr. (c) Cells were treated with 40 μg/ml angelicin for 1.5 hr, irradiated with 340-nm UV for 13 min at a fluence rate of 23.5 J/m² per sec on 4 ml *Drosophila* Ringer solution, and incubated for a further 6 hr in a fresh medium. (f) Cells were irradiated at a dose of 90 J/m², then 6 ml medium added and incubated for 6 hr. (j) Cells were exposed for 37°C for 2 hr.

On exposure of *Drosophila* cells to temperatures (28–39°C) above normal (25°C), a small set of genes are newly activated, whereas most normally active genes are inactivated. The maximum response is observed at 36–37°C. Expression of HSPs is controlled at both transcriptional and posttranscriptional levels. Most HSPs are also synthesized constitutively at least in some tissues and during specific developmental stages (Lindquist, 1986; Ashburner and Bonner, 1979).

How are the HS genes activated by HS? HS genes contain multimers of a 5-bp sequence (NGAAN) referred to as the heat shock element (HSE) that is arranged in alternating orientation in the 5'-flanking region. The HS-inducible transcription factor (HSF) binds to HSE, and HS genes are activated (Xiao and Lis, 1988; Amin *et al.*, 1988). Various factors may be responsible for activation of the HSF genes but precise mechanisms are unknown.

Functional aspects of HSPs had been in the dark for a long time. Possible roles of some of the HSPs were proposed by Pelham (1986). The HSP 70 family and possibly the HSP 90 family are involved in the assembly and disassembly of proteins and protein-containing structures, coupled with ATP hydrolysis. Taken together with other examples, the concept of molecular chaperones has been proposed (Ellis, 1993). The molecular chaperones are defined as a class of unrelated families of proteins that assist the correct noncovalent assembly of other polypeptide-containing structures *in vivo*. These chaperone molecules themselves are not components of the assembled structures when they are performing their normal biological functions. The concept includes the proposed roles of some HSPs. The term *protein assembly* is used in a wide sense: it covers functions to repair or destroy proteins damaged by stresses. The HSP 70 family and HSP 83 have in fact been shown to work as molecular chaperones in *Drosophila*. Small HSPs are related to each other and to α-crystallin. They may have chaperone activity, but their functions remain unclear. For recent reviews of particular aspects of HSP molecular chaperones, see Parsell and Lindquist (1993) and Hendrick and Hartl (1993).

3.2.2. *Effect of Hydrogen Peroxide on HSP Gene Activation*

The HSP genes are activated by various agents, as can be seen clearly in induced puffs on the polytene chromosomes (Ashburner and Bonner, 1979). The effect of DNA-damaging agents (H_2O_2, UV and γ-ray irradiations) on the expression of major HSP genes was investigated by using the cloned HSP DNAs (Love *et al.*, 1986). The nuclei were isolated from untreated and treated Schneider's line 2 tissue culture cells, and nuclear transcripts were labeled *in vitro* by RNA polymerase II with ^{32}P-UTP under conditions to permit an elongation of previously initiated RNA chains. Labeled runoff transcripts were hybridized to appropriate amounts of cloned cDNAs of the HSP (83, 70, 68, 27, 26, 23, and 22 kDa) genes immobilized on filters. Love *et al.* (1986) showed that HSP transcription was enhanced by H_2O_2 but not by UV light or γ rays. Genes for small HSPs were activated more efficiently than those for larger ones. Furthermore, pretreatment of cells with low doses (1 mM, 3 days) of H_2O_2 or with HS (37°C, 1 hr) subsequent to exposure to the high dose (50 mM, 1 hr) of H_2O_2, resulted in about twofold increases in the production of HSP (83, 70, 26, and 22) transcripts, compared with the induction observed following the high dose treatment only. Induction of the HSP genes enhanced toxicity of high dose exposure, as assessed by impairment in uridine incorporation and cell proliferation. This result is contrary to the observation that mild preheat treatment enhanced the tolerance to severe HS. Another group investigated the relationship between cell survival and HSP expression (Tomasovic and Koval, 1985). Mild HS (33°C, 30 min) led to thermotolerance on 1- to 6-hr incubation at 28°C, whereas stronger HS (37°C, 30 min) did not lead to thermotolerance over the 6-hr interval, even though the latter HS induced more HSPs. The pretreatment (HS and H_2O_2) in the H_2O_2 experiments mentioned above might have been too severe to the cells. Taken together, it can be concluded that pretreatment with low doses of toxic agents or HS alters the tolerance of cells. In the system described in the previous section (3.1), H_2O_2 induced some HSPs and UV showed little induction of HSPs (Table I).

3.3. Identification of the mRNAs Induced by MMS

3.3.1. Isolation of cDNA Clones from Inducible mRNAs

mRNAs were purified from the untreated cells and the MMS-treated cells. cDNAs labeled with ^{32}P-CTP were prepared from both mRNA fractions separately. The ^{32}P-cDNAs corresponding to induced mRNAs were enriched by the procedure with two cycles of hybridization as described in Section 2.2.2. Both enriched and untreated cDNAs were used as probes for plaque hybridization. Two hundred twenty plaques, with signals to the enriched cDNAs stronger than to the untreated cDNAs, were isolated from 2×10^4 plaques. This mutagen-sensitive gene library was then examined as below.

HSPs were induced by MMS treatment (Table I). The cDNAs coding for HSP 83k, 70k, 26k, and either 21k or 23k, and 22k (kindly provided by Drs. J. T. Lis and E. A. Craig) were used as probes for plaque hybridization. Forty percent of the clones in the above mutagen-sensitive gene library hybridized to one of these HSP probes.

3.3.2. Induced Proteins

The cDNA clones that hybridized to major HSP cDNAs were eliminated from further analysis, as our purpose was to analyze genes specific to the MMS treatment. The remaining clones were treated as follows: Several clones that showed positive signals ranging from strong to weak in the first plaque hybridization (Section 3.3.1) were selected as representatives. Insert DNA fragments were prepared from each of these clones and examined by either a Northern blot or spot test hybridization to determine whether they had the sequence that was newly transcribed following treatment with MMS. The insert DNAs confirmed as such were then hybridized to the mutagen sensitive gene library. The clones that showed positive signals against each of the representatives mentioned above constitute the same group. Each group is named after the representative clone number. Several clones that did not show appreciable signals in the above test were also selected and examined.

The nucleotide sequence of the insert DNA of one clone from each group was determined and examined for homology with known sequences in the GenBank DNA database by using the FASTA program. For the DNAs that had not previously been identified in *Drosophila*, their cytological locations were determined by *in situ* hybridization of the cloned DNA to the Oregon-R (*Drosophila* wild type) polytene chromosomes using biotinylated DNA probes. One typical map is shown in Fig. 2.

The results of the homology search and cytological locations are summarized in Table II (Akaboshi *et al.*, 1994b). The previously identified genes with which the homology is greatest are listed. When the homologues are found in several organisms, the human homologue is represented.

Glutathione S-transferase (GST) (group 35) is a family of multifunctional proteins that conjugate glutathione on the sulfur atom of the cysteine to various electrophiles. GST also binds a variety of hydrophobic compounds. Thus, GST participates in detoxification of xenobiotics, in drug metabolism, and in protection against peroxidative damage (review: Tsuchida and Sato, 1992). Our GST gene is homologous to the GST D genes (Toung *et al.*, 1993). MMS was the strongest inducer among the agents examined. GST mRNA from the untreated cells was 0.8 kb long on Northern blots, which is approximately the same size as that of the inserts of clones isolated by Toung *et al.* (1990). Two newly induced bands (1.6 and 2.8 kb) were detected in addition to the increased strength of the 0.8-kb band by Northern hybridization, following treatment for 3 or 6 hr with 0.025% MMS. Nucleotide sequences of 1.6- and 1.8-kb inserts of the two clones (group 35) revealed that the 70-bp sequence and its variants were repeated.

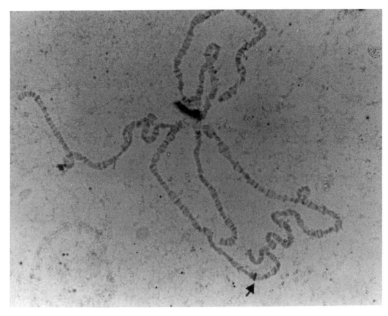

FIGURE 2. *In situ* hybridization to *Drosophila* polytene chromosomes. The arrow indicates the position, 55C on chromosome 2R, at which #53 cDNA hybridized.

These repeated sequences had no homology with the GST 1-1 gene (Toung *et al.*, 1990). Treatment of cells with EMS, UV, and 4-NQO induced the GST gene, whereas treatment with MMC and H_2O_2 also induced this gene, but only slightly. HS and TPA were without effect. New bands (1.6 and 2.8 kb) as mentioned above, however, were not seen with the other inducers examined. Potential candidates of mutation occupying the same cytological position, 86E/F (Table II), were searched for in a comprehensive list of mutants (Lindsley and Zimm, 1992). A possible candidate (*mus309*) was found, in that larval survival is hypersensitive to MMS and nitrogen mustard.

Ubiquitin (group 42) has one of the most conserved sequences throughout evolution. It is known to associate with the short-lived and nonnative proteins that mark for degradation (review: Finley and Chau, 1991). The *Drosophila* genome possesses *ubi-m* and *ubi-p* loci. The latter is composed of tandem repeats (in wild-type Canton S, 28 repeats) of a 228-bp ubiquitin coding unit that

TABLE II
cDNAs for mRNAs Induced by MMS Treatment

Clone name	Insert (kb)	Cytological location	Comment
23	1.9	4F, 48D/E	Ribosomal protein S11 (HUMRPS11, 65% in 197 bp)
35	1.6	86E/F	D. m. glutathione *S*-transferase 1-1 (DROGST, 78% in 451 bp)
42	1.2		Ubiquitin
91	1.4	42C/D	Human *gadd45*[a]
128	1.9	98F	Human cystic fibrosis TR mRNA (HUMCFTRM, 53% in 1046 bp)
145	0.65	8B/C	Pig moesin B (PIGMOESB, 55% in 205 bp)
214			D. m. dorsal

Clones with no previously known sequences (cytological location)
4 (62C); 43 (91C/D); 53 (55C); 82 (22F); 158 (7A); 163 (77E/F); BS1[b] (29C); BS3[b] (46B/C); BX203[b] (21A); 217 (6C, 8E, 29A, 43C, 49D, 55A, 59F, 85D, 88E, 91A)

[a]See text.
[b]These MMS-inducible clones were found in a separate experiment.

corresponds to a 76-amino-acid (aa) polypeptide (Lee *et al.*, 1988). Ubiquitin mRNA showed a clear band at 4.2-kb with smears until 0.8 kb in a Northern blot. MMS was the most effective inducer of ubiquitin among the agents examined. H_2O_2 and EMS were only slightly effective. UV and TPA, on the contrary, repressed the expression. With HS, little of the 4.2 kb band was detectable but small molecular messages (smears) showed an increase.

Group 91 (Table II) showed homology with the human *gadd45* (growth arrest and DNA damage inducible) gene with the aa alignment, BLASTX. The *Drosophila* homologue, which codes for a 163-aa polynucleotide, showed 33% identity and 66% possibility over 82 aa (partially 40 and 72%, respectively, over 37 aa) with the human protein (165 aa). The *gadd45* genes were induced in response to DNA-damaging agents and a variety of other growth cessation signals such as reduction in the serum level and depletion of nutrients in the culture medium in rodent and human cells (review: Fornace, 1992). These treatments cause accumulation of cells in the G_0 phase. The *gadd45* gene, together with four other *gadd* genes, are induced by UV, MMS, and x rays, but not by HS (Fornace, 1992; Papathanasiou *et al.*, 1991). *Drosophila* homologues for the four genes have not been isolated. The fact that the *gadd45* homologue has been detected as an inducible gene by MMS, but not by HS, in *Drosophila*, suggests that this gene and its expression are highly conserved throughout evolution in higher eukaryotes.

Group 128 (Table II) showed a very high homology (53% in 1046 bp) to the human cystic fibrosis (CF) transmembrane conductance regulator (CFTR) gene (Table II) and a similarly high homology (52% in 733 bp) to the *D. melanogaster* P-glycoprotein (Pgp) gene. Both the CFTR and Pgp are transmembrane proteins and are members of a superfamily of ATP-binding cassette or traffic ATPases (reviews: Sferra and Collins, 1993; Doige and Ames, 1993). CF is a lethal autosomal recessive disease with multi-organ system disorders, arising from a single biochemical abnormality. CFTR is a chloride channel. It possibly participates in other functions (e.g., other ion channels and a transporter of large molecules) as well. Hydrolysis of ATP may also be involved in signaling pathways.

Group 145 (Table II) showed a very high homology with the pig moesin B gene (Table II). Moesin B, radixin, and erzin are related members of a family of submembrane cytoskeletal proteins (Lankes *et al.*, unpublished). Their function, however, remains unclear.

The sequence of clone 214 was found to have homology with the dorsal (*dl*) gene of *D. melanogaster*. The Dorsal (Dl) protein is known to act as the maternal morphogen that controls the dorso-ventral (DV) polarity through its selective nuclear import in *Drosophila* embryos. The DV pattern is established by a process involving signal transduction through the interaction of proteins coded by a set of maternal and zygotic genes. Their ultimate task is to create an intranuclear gradient of the Dl protein. The *dl* gene is homologous to the vertebrate proto-oncogene, *c-rel*, and to the *NF-κB* genes. The *NF-κB* gene is γ ray inducible (Fornace, 1992). They have about 50% identical aa over a sequence of approximately 300 aa in the amino-terminal region (reviews: Siebenlist *et al.*, 1994; Steward and Govind, 1993; Schmitz *et al.*, 1991). The c-Rel/NF-κB protein is a transcription factor and participates in various kinds of cellular reactions such as acute-phase responses and lymphocyte differentiation. The Dl protein is found to be ·a sequence-specific DNA-binding protein within a repressor element of the *zen* gene (Ip *et al.*, 1991), and functions also as an activator for other genes (*twist* and *snail*). The *zen*, *twist*, and *snail* genes are all involved in DV pattern formation. In our present experimental system, the expression of the *dl* gene was enhanced at 6-hr treatment but not at 3 hr. This observation might be explained by a mechanism similar to the induction of the NF-κB protein, in that the initial induction does not require *de novo* protein synthesis and involves dissociation of a latent cytoplasmic complex with inhibitor protein (I-κB) and the nuclear import of the NF-κB protein (Schmitz *et al.*, 1991). Similarly, the Dl protein is released from the cytoplasmic inhibitor cactus and enters the nuclei forming a gradient along the DV pattern (Steward and Govind, 1993). In addition to this role, recent data from cloning the genes induced by injury or infection have indicated that the *dl* and related genes may play an important role in the "immune response" (see next section).

3.3.3. Immune Response

When insects including *Drosophila* are subjected to injury or to the experimental injection of bacteria, a large number of antibacterial peptides are synthesized rapidly and accumulate transiently (reviews: Cociancich *et al.*, 1994; Hultmark, 1993). This is a so-called "immune response" in insects, evocative of the acute phase response in mammals. Some of these "immunity" genes are expressed during normal development. The upstream regions of all of these genes contain sequence motifs related to the κB motif to which the NF-κB protein binds. The main site for the synthesis of the peptides is the fat body, the functional homologue of the mammalian liver that is the main tissue for acute-phase response. Because the fat body does not interact directly with bacteria, signals must be transmitted from the infected site to cells in the fat body. These similarities to the NF-κB suggested that immune reaction might be mediated by transcription factors related to NF-κB. Indeed, the Dl and Dif (Dorsal-related immunity factor) proteins have been suggested to participate in the immune response (Lemaitre *et al.*, 1995; Ip *et al.*, 1993). When a larva is injected with bacteria or damaged by manual puncture, the Dif protein accumulates rapidly in the nuclei of fat body cells. It does not appear to participate in the DV patterning (Ip *et al.*, 1993). The *dl* message increases in the fat body on injury. The Dl protein is involved in the activation of the diptericin gene, one of the inducible antibacterial genes (Lemaitre *et al.*, 1995). The nuclear localization of Dl is controlled by the same signaling pathway as in DV patterning. Expression of the diptericin gene, however, can occur in the absence of Dl. Thus, the process of immune response appears not to be a simple one. Some insect cell lines are found to display an inducible immune response (Hultmark, 1993). These cell culture systems may provide a means to clarify the mechanisms involved in immune reactions at the genetic and molecular levels, as identification and characterization of the genes related to the *dl* gene can be made (and indeed are being made) more easily in *Drosophila* than in mammals. Information from *Drosophila* may help us understand the mammalian systems.

3.3.4. Other Aspects of the MMS Response

During our analysis of cDNAs from MMS-inducible mRNA, we encountered a variety of mRNA species, in addition to those described in the previous section. Some of them seem to be worth mentioning here.

The insert DNA of the clones of some of the groups in Table II contained partially varied sequences. One (group 53) was found to contain sequences with alternative splicing sites, based on a comparison of sequences of the cDNAs with the cloned genomic DNA sequence. In group 217, the nucleotide sequences of insert DNAs (4.9, 3.9, and 1.9 kb) from three clones were determined. They have partially different sequences and the junction site varies. As the insert DNA of clone 217 was mapped at 10 sites on the polytene chromosomes (Table II), a transposonlike sequence may possibly be involved.

One clone in group 128 had a sequence (390 bp) homologous to human mRNA (386 bp) encoding a ribosomal protein (rp) (homologue of yeast rp L44) through poly(T)$_{44}$. As the two fragments were in an opposite direction, they could serve as a protein-coding sequence. The rp homologue was also found to be highly homologous (93%) to suffix sequences (Tchurikov *et al.*, 1986). The suffix sequences (265 bp) are located at the 3'-ends of various mRNAs, but they seem not to serve as messages because many termination codons appear in all three frames. Our clone, on the other hand, contained no termination codons. The L44 rp homologue was not inducible by MMS in our experiments.

At least one clone of group 23 had a homologous sequence to the rp S11 gene in the 2.3-kb sequence (Table II). This sequence was induced early (at 3 hr) but not late (at 6 hr) following the MMS treatment.

Several clones were found to have apparently conjugated mRNAs. One clone from group 214 contained 1.7-kb dorsal (*dl*) and 0.1-kb β-tubulin sequences. A ubiquitin sequence was also isolated as a carrier of unknown gene(s) with tandem repeating sequences. One of them was found in group 53.

The transposon, TN1731, that mimics induction of prophage in bacteria (review: Walker, 1984) and of virus sequences in mammals (Table III) was induced.

3.3.5. Overview of MMS Response

No genes that participate in such processes as DNA repair and recombination have been found among the genes induced by MMS. This, however, should not be taken to mean that the DNA-repair enzyme genes are not induced by MMS, because the number of the sequences that have been analyzed is too small and also because all of the repair genes have not yet been identified.

The major HSP, GST, and ubiquitin genes were activated. The proteins from these three classes of genes are known to participate in degradation, exclusion, and repair of modified or denatured proteins. It is possible that they participate in other activities as well.

The CFTR homologues are all membrane-associated proteins. The group 53 clones have sequences containing the same nucleotide motifs, and thus they possibly also participate in some signaling pathways.

The *dl* message was induced by MMS a little later in time after treatment in comparison with the early expressed genes. The distribution of the transcription factor between the nucleus and cytoplasm may be important in response to stress as suggested by the observations in DV patterning during normal development. The activation of the *dl* gene might suggest that the DNA-damaging agents such as MMS activate some of the genes related to the "immune" reaction in a similar manner. Some of the clones that are not yet thoroughly studied (Table II) may contain sequences for such transcription factors (Akaboshi, unpublished).

The *gadd45* homologue might be associated with growth arrest as in mammalian cells.

Taken together, the cellular response against the MMS treatment appears to be controlled by multiple systems. Several types

TABLE III
Genes Inducible by DNA-Damaging Agents in Mammalian Cells

Oncogene	DNA repair	Viruses	Refs.
c-fos, c-jun, jun-B, jun-D, TRK, p53, c-myc, c-H-ras, MDM	MGMT, MAG, β-polymerase	HIV-1, Mo-MUSV, SV40, VL30, RaLV	Herrlich *et al.* (1992), Fornace (1992) Chen *et al.* (1994), Price and Park (1994)
	Other genes coding for proteins		
NF-κB, Pgp, α-interferon, IL-1, bFGF, EGF receptor, EGR1, TNFα, PKC, *sprI, sprII, spr2-1*, RP-2, RP-8, metallothioneins, urokinase, ornithine decarboxylase, collagenase, heme oxygenase, DDI-class I, DDI-class II, MHC class I, α-tubulin, β, γ-actin, keratins, invariant chain, GAPDH, DNA-binding proteins, DNA ligase, collagens, Myd118			Herrlich *et al.* (1992), Fornace (1992)
RNA-binding protein			Carrier *et al.* (1994)

of clones containing sequences combining different mRNA species have been observed. Such complex messages might be one of the most attractive experimental materials for studies on the expression of genes and their roles in the inducible response.

4. COMPARISON WITH OTHER EUKARYOTIC CELLS

The effects of DNA damage have been investigated intensively in mammalian cells especially in relation to cancer (reviews: Herrlich *et al.*, 1992; Fornace, 1992). DNA-damaging agents elicit complex responses, including growth rate changes and the induction of a variety of genes. Names of the inducible genes so far identified are listed in Table III. Induced genes among higher eukaryotes including *Drosophila* may be compared. Genes identified after the above-mentioned two reviews are also included with references. In most of the reported cases, induction of the genes was noted in individual experiments rather than in systematic experiments such as by using the subtraction method (Fornace, 1992). The genes appear to be associated with diverse cellular processes. However, it must be emphasized that most of the genes were inducible only by particular means. Note also that the induction has been studied in many different ways utilizing a number of organisms, cell types, kinds and doses of inducers, and kinds of inducible materials (e.g., RNA, DNA, protein).

The direct effects of DNA-damaging agents on eukaryotic cells may be the induction of DNA repair-related genes as is known to occur in prokaryotes, and especially well studied in *E. coli* (Walker, 1984). In *Saccharomyces cerevisiae*, a lower eukaryote, some DNA repair genes have been shown to be induced on damaging DNA (reviews: Prakash *et al.*, 1993; Friedberg, 1988). In higher eukaryotes, however, only a few examples have been reported (Table III). The N^3-methyladenine glycosylase (MAG) and O^6-methylguanine-DNA methyltransferase (MGMT) genes are induced by γ rays, and by x rays, γ rays, and UV, respectively, as well as by alkylating agents, although these enzymes are presumably involved in the repair of non-UV-type damages. The activation of the MAG and MGMT genes occurs only in rat hepatoma cells, and rat and mouse hepatoma cells, respectively. The MGMT protein was also induced by HS. The induction of β-polymerase has been reported only in CHO cells. The mRNA level of β-polymerase is elevated after treatment with MNNG, MMS, and H_2O_2, all of which are known to produce single-base damages, but this increase in the β-polymerase mRNA level results in no significant increase in the β-polymerase activity level in the cell extract following treatment with MNNG and MMS. Thus, at present, strong evidence for the induction of DNA repair genes seems to be lacking in higher eukaryotes.

Several types of DNA repair systems have been shown to be operative in higher eukaryotes (reviews: Modrich, 1991; Bohr *et al.*, 1989). Different genes within the same cell are repaired with different efficiencies. Active genes, such as essential genes and inducible genes when the gene is transcriptionally active, are preferentially repaired (Bohr *et al.*, 1989). Studies on the nonhomogeneous repair of the UV-irradiated dehydrofolate reductase (DHFR) gene in rodent and human cells (Lommel *et al.*, 1995; Bohr *et al.*, 1989) have revealed that the transcribed DNA strands are repaired preferentially in comparison with the nontranscribed strands, as has also been demonstrated in bacteria and yeast. In higher eukaryotes, DNA is rigidly associated with proteins to form chromatin most of the time and thus may be resistant to attack by various agents. Then it may be that constitutive enzymes are sufficient for the repair of damaged DNA. Alternatively, enzymes as yet unidentified may reveal such inducible function.

The regulatory systems that are relevant to the expression of DNA repair genes must also have developed during evolution. We may take the *recA* gene as an example. In bacteria, the *recA* gene, indispensable in the DNA repair process, is always induced by DNA-damaging agents (Walker, 1984). Two *recA* homologues (*DMC1* and *RAD51*) are known in S. *cerevisiae*. The *RAD51* transcript is induced by MMS treatment, and the mutant strains (RAD51⁻) are sensitive to MMS (Shinohara *et al.*, 1992). The *DMC1* mutants, on the other hand, are not sensitive to MMS (Bishop *et al.*, 1992). The homologous gene (DMR) in *Drosophila* (Akaboshi *et al.*, 1994a) is not induced by the MMS treatment, but actually is repressed.

In *E. coli*, some genes, such as the *sulA* gene which causes the long filamentous morphology following the expression of "SOS" genes, are induced by DNA-damaging agents (Walker, 1984). The SulA protein apparently acts as an inhibitor of cell division. Induction of the genes that have no function in DNA repair has been largely neglected as irrelevant or unclear aspects of "SOS functions." Such genes, however, may have developed as primary inducible responses to varying intracellular and intercellular events in organisms.

As for the inducible genes associated with various functions other than DNA repair, the similarities seem to exist between *Drosophila* (Table II) and mammals (Table III), although only a small number of genes are listed at present in *Drosophila*. HSP and ubiquitin genes are inducible both in mammals and in *Drosophila*. Transmembrane proteins (CFTR and Pgp) might be involved in signaling pathways. Both the *gadd45* and *NF-κB/dorsal* homologues are inducible. The GST gene in *Drosophila* is inducible. Inducibility in the GST genes in mammals remains to be examined. Some of the forms such as rat GST-P and human GST-π have been used as reliable preneoplastic or neoplastic marker enzymes, and the mechanism of their expression has been and is being studied in relation to oncogene activation (Tsuchida and Sato, 1992). Roles of the gene products, however, are little understood. Identification of the genes and isolation of the homologues suggest that similar pathways may be working in these organisms.

5. DISCUSSION

DNA-damaging agents may directly interact with DNA to cause DNA damage, or they may elicit a cellular response by a mode independent of their DNA-damaging action. As little generalized evidence has been obtained for the induction of DNA repair genes in higher eukaryotes, it is possible that DNA repair genes may not constitute the main category for induction.

Induction of a variety of genes by DNA-damaging agents has been revealed (Tables II and III). Some of these genes are commonly induced by other kinds of stresses, such as HS, cell injury, inflammation, and immune responses, and a few are even expressed as normal cellular functions (growth regulation and cell differentiation). A number of similarities among the "stress reactions" are now only beginning to emerge. Cellular response to various agents appears to be highly conserved among various organisms. At present, however, we have to be careful to recognize that just because homologous genes are induced does not necessarily mean that they have similar functions. The organism may respond to each stress in a case-by-case manner, as the genes do not appear to be regulated by a single mechanism. Still we would want to see if a simple common mechanism exists to explain diverse reactions, e.g., the DNA structure is determined by the sequence itself, but may be modified by binding with other substances.

Another way may be to examine systematically whether many more inducible genes exist. Although we already have many inducible genes in hand to study, we still do not know if individual sets of genes induced play an agent-specific role. Some genes might be induced only as a sign for an emergency. Organisms may have a common mechanism as a primitive host-defense at the basal level, to alarm cells such as in the nervous system. The exposure to stresses may lead to the transient pose or cessation of normal activities locally or totally, and such might be an initial signal for the subsequent induction of a variety of genes.

Living organisms must respond to a stress by using the whole body even when the stress is a very small one. In addition to the induction of particular sets of genes, modulating systems of the gene products may also be important as well. The stress response should be considered as a total system. We hope studies on the expression of stress genes might offer a useful barometer to monitor the status of the environment by using living cells.

ACKNOWLEDGMENTS. We thank Dr. Kugao Oishi for discussion and critical reading of the manuscript. We are grateful to our colleagues in our laboratories.

REFERENCES

Akaboshi, E., and Howard-Flanders, P., 1989, Proteins induced by DNA-damaging agents in cultured *Drosophila* cells, *Mutat. Res.* **227**:1–6.

Akaboshi, E., Inoue, Y., and Ryo, H., 1994a, Cloning of the cDNA and genomic DNA that correspond to the *recA*-like gene of *Drosophila melanogaster, Jpn. J. Genet.* **69**:663–670.

Akaboshi, E., Inoue, Y., Ryo, H., and Yamamoto, M. T., 1994b, Cloning and mapping of genes activated by MMS treatment in *Drosophila* cultured cells, *Dros. Inf. Serv.* **75**:149.

Amin, J., Ananthan, J., and Voellmy, R., 1988, Key features of heat shock regulatory elements, *Mol. Cell. Biol.* **8**:3761–3769.

Ashburner, M., and Bonner, J. J., 1979, The induction of gene activity in *Drosophila* by heat shock, *Cell* **17**:241–254.

Bishop, D. K., Park, D., Xu, L., and Kleckner, N., 1992, *DMC1*: A meiosis-specific yeast homolog of *E. coli recA* required for recombination, synaptonemal complex formation, and cell cycle progression, *Cell* **69**:439–456.

Bohr, V. A., Evans, M. K., and Fornace, A. J., Jr., 1989, Biology of disease: DNA repair and its pathogenetic implications, *Lab. Invest.* **61**:143–161.

Carrier, F., Gatignol, A., Hollander, M. C., Jeang, K.-T., and Fornace, A. J., Jr., 1994, Induction of RNA-binding proteins in mammalian cells by DNA-damaging agents, *Proc. Natl. Acad. Sci. USA* **91**:1554–1558.

Chen, C.-Y., Oliner, J. D., Zhan, Q., Fornace, A. J., Jr., and Vogelstein, B., 1994, Interactions between p53 and MDM2 in a mammalian cell cycle, *Proc. Natl. Acad. Sci. USA* **91**:2684–2688.

Cociancich, S., Bulet, P., Hetru, C., and Hoffmann, J. A., 1994, The inducible antibacterial peptides of insects, *Parasitol. Today* **10**:132–139.

Doige, C. A., and Ames, G. F.-L., 1993, ATP-dependent transport systems in bacteria and humans: Relevance to cystic fibrosis and multidrug resistance, *Annu. Rev. Microbiol.* **47**:291–319.

Ellis, R. J., 1993, The general concept of molecular chaperones, *Philos. Trans. R. Soc. London Ser. B* **339**:257–261.

Engels, W. R., Preston, C. R., Thompson, P., and Eggleston, W. B., 1986, *In situ* hybridization to *Drosophila* salivary chromosomes with biotinylated DNA probes and alkaline phosphatase, *Focus* **8**:6–8.

Finley, D., and Chau, V., 1991, Ubiquitination, *Annu. Rev. Cell Biol.* **7**:25–69.

Fornace, A. J., Jr., 1992, Mammalian genes induced by radiation; activation of genes associated with growth control, *Annu. Rev. Genet.* **26**:507–526.

Friedberg, E. C., 1988, Deoxyribonucleic acid repair in the yeast, *Microbiol. Rev.* **52**:70–102.

Hendrick, J. P., and Hartl, F.-U., 1993, Molecular chaperone functions of heat-shock proteins, *Annu. Rev. Biochem.* **62**:349–384.

Herrlich, P., Ponta, H., and Rahmsdorf, H. J., 1992, DNA damage-induced gene expression: Signal transduction and relation to growth factor signaling, *Rev. Physiol. Biochem. Pharmacol.* **119:**187–223.

Hultmark, D., 1993, Immune reactions in *Drosophila* and other insects: A model for innate immunity, *Trends Genet.* **9:**178–182.

Ip, Y. T., Kraut, R., Levine, M., and Rushlow, C. A., 1991, The *dorsal* morphogen is a sequence-specific DNA-binding protein that interacts with a long-range repression element in *Drosophila*, *Cell* **64:**439–446.

Ip, Y. T., Reach, M., Engstrom, Y., Kadalayil, L., Cai, H., Gonzalez-Crespo, S., Tatei, K., and Levine, M., 1993, *Dif*, a *dorsal*-related gene that mediates an immune response in *Drosophila*, *Cell* **75:**753–763.

Kelley, P. M., and Schlesinger, M. J., 1978, The effect of amino acid analogues and heat shock on gene expression in chicken embryo fibroblasts, *Cell* **15:**1277–1286.

Lee, H., Simon, J. A., and Lis, J. T., 1988, Structure and expression of ubiquitin genes of *Drosophila melanogaster*, *Mol. Cell. Biol.* **8:**4727–4735.

Lemaitre, B., Meister, M., Govind, S., Georgel, P., Steward, S., Reichhart, J.-M., and Hoffmann, J. A., 1995, Functional analysis and regulation of nuclear import of *dorsal* during the immune response in *Drosophila*, *EMBO J.* **14:** 536–545.

Lindquist, S., 1986, The heat-shock response, *Annu. Rev. Biochem.* **55:**1151–1191.

Lindsley, D. L., and Zimm, G. G., 1992, *The Genome of Drosophila melanogaster*, Academic Press, San Diego.

Lommel, L., Carswell-Crumpton, C., and Hanawalt, P. C., 1995, Preferential repair of the transcribed DNA strand in the dihydrofolate reductase gene throughout the cell cycle in UV-irradiated human cells, *Mutat. Res.* **336:**181–192.

Love, J. D., Vivino, A. A., and Minton, K. W., 1986, Hydrogen peroxide toxicity may be enhanced by heat shock gene induction in *Drosophila*, *J. Cell. Physiol.* **126:** 60–68.

Modrich, P., 1991, Mechanisms and biological effects of mismatch repair, *Annu. Rev. Genet.* **25:**229–253.

Nover, L., 1991, Structure of eukaryotic heat shock genes, in: *Heat Shock Response* (L. Nover, ed.), CRC Press, Boca Raton, FL, pp. 129–150.

Papathanasiou, M. A., Kerr, N. C. K., Robbins, J. H., McBride, O. W., Alamo, I., Jr., Barrett, S. F., Hickson, I. D., and Fornace, A. J., Jr., 1991, Induction by ionizing radiation of the *gadd45* gene in cultured human cells: lack of mediation by protein kinase C, *Mol. Cell. Biol.* **11:**1009–1016.

Parsell, D. A., and Lindquist, S., 1993, The function of heat-shock proteins in stress tolerance: Degradation and reactivation of damaged proteins, *Annu. Rev. Genet.* **27:**437–496.

Pelham, H. R. B., 1986, Speculations on the functions of the major heat shock and glucose-regulated proteins, *Cell* **46:**959–961.

Prakash, S., Sung, P., and Prakash, L., 1993, DNA repair genes and proteins of *Saccharomyces cerevisiae*, *Annu. Rev. Genet.* **27:**33–70.

Price, B. D., and Park, S. J., 1994, DNA damage increases the levels of *MDM2* messenger RNA in wtp53 human cells, *Cancer Res.* **54:**896–899.

Ritossa, F., 1962, A new puffing pattern induced by a temperature shock and DNP in *Drosophila*, *Experientia* **18:**571–573.

Schmitz, M. L., Henkel, T., and Baeuerle, P. A., 1991, Proteins controlling the nuclear uptake of NF-κB, rel and dorsal, *Trends Cell Biol.* **1:**130–137.

Sferra, T. J., and Collins, F. S., 1993, The molecular biology of cystic fibrosis, *Annu. Rev. Med.* **44:**133–144.

Shinohara, A., Ogawa, H., and Ogawa, T., 1992, Rad51 protein involved in repair and recombination in S. *cerevisiae* is a RecA-like protein, *Cell* **69:**457–470.

Siebenlist, U., Franzoso, G., and Brown, K., 1994, Structure, regulation and function of NF-κB, *Annu. Rev. Cell Biol.* **10:**405–455.

Steward, R., and Govind, S., 1993, Dorsal–ventral polarity in the *Drosophila* embryo, *Curr. Opin. Genet. Dev.* **3:**556–561.

Tchurikov, N. A., Ebralidze, A. K., and Georgiev, G. P., 1986, The suffix sequence is involved in processing the 3′ ends of different mRNAs in *Drosophila melanogaster*, *EMBO J.* **5:**2341–2347.

Tomasovic, S. P., and Koval, T. M., 1985, Relationship between cell survival and heat-stress protein synthesis in a *Drosophila* cell line, *Int. J. Radiat. Biol.* **48:** 635–650.

Toung, Y.-P. S., Hsieh, T.-S., and Tu, C.-P. D., 1990, *Drosophila* glutathione S-transferase 1-1 shares a region of sequence homology with the maize glutathione S-transferase III, *Proc. Natl. Acad. Sci. USA* **87:**31–35.

Toung, Y.-P. S., Hsieh, T.-S., and Tu, C.-P. D., 1993, The glutathione S-transferase D genes. A divergently organized, intronless gene family in *Drosophila melanogaster*, *J. Biol. Chem.* **268:**9737–9746.

Tsuchida, S., and Sato, K., 1992, Glutathione transferases and cancer, *Crit. Rev. Biochem. Mol. Biol.* **27:**337–384.

Walker, G. C., 1984, Mutagenesis and inducible responses to deoxyribonucleic acid damage in *Escherichia coli*, *Microbiol. Rev.* **48:**60–93.

Welcher, A. A., Torres, A. R., and Ward, D. C., 1986, Selective enrichment of specific DNA, cDNA and RNA sequences using biotinylated probes, avidin and copper-chelate agarose, *Nucleic Acids Res.* **14:**10027–10044.

Wilder, R. L., 1995, Neuroendocrine–immune system interactions and autoimmunity, *Annu. Rev. Immunol.* **13:**307–338.

Xiao, H., and Lis, J. T., 1988, Germline transformation used to define key features of heat-shock response elements, *Science* **239:**1139–1142.

4

Stress Resistance in Lepidopteran Insect Cells

THOMAS M. KOVAL

1. INTRODUCTION

Insects represent the most populous group of animals on Earth. Over 75% (over 932,000 out of approximately 1,250,000) of all described animal species are insects. Insects occupy and play an important role in virtually every terrestrial ecosystem and many aquatic ecosystems. From worldwide economic and human health perspectives, approximately 20% of all crops are lost because of insects and one in six individuals has an insect-borne illness. Clearly, from both prevalence and relevance points of view, insects merit our attention and study.

Lepidopteran insects (moths and butterflies) have an ancestry that dates to the Mesozoic Era (Labandeira and Sepkoski, 1993). During the past 200 million years, these organisms have been successful at resisting numerous predators as well as exposures to a continuous barrage of physical, chemical, and biological agents. These agents include nature's own hazards such as the

THOMAS M. KOVAL • National Council on Radiation Protection and Measurements, Bethesda, Maryland 20814.

Stress-Inducible Processes in Higher Eukaryotic Cells, edited by Koval. Plenum Press, New York, 1997.

ultraviolet component of sunlight and hazardous natural chemicals in plants that serve to deter ingestion by insects, and a wide variety of synthetic and other agents applied by humans in eradication efforts. Given the survival and diversity of lepidopteran insects in the face of these various stress exposures over this enormous period of time, lepidopteran cells would seem to be a model system for examining stress responses.

2. COMPARATIVE RESISTANCE

2.1. X and Gamma Rays

Compared with other organisms, insects exhibit a pronounced inherent resistance to ionizing radiation. In general, adult insects are considered to be at least 100 times less sensitive to such radiation than are vertebrates (O'Brien and Wolfe, 1964). The range of lethal exposures to ionizing radiation is generally between 2 and 9 Gy for most mammals (Bond *et al.*, 1965) and 1 and 1000 Gy for most adult insects (Ducoff, 1972; Casarett, 1968).

Cells from insects retain this pronounced intrinsic resistance to ionizing radiation even when proliferating *in vitro* (Koval *et al.*, 1975). This finding points toward an inherent cellular basis for the pronounced resistance. Although they possess marked relative resistance, insect cells nevertheless display typical radiation-induced morphological alterations, albeit at higher dose levels than other cell types. Studies involving cells from several species of Lepidoptera demonstrated that extraordinary radiation resistance is typical of this insect order, generally being about 100 times more resistant than mammalian cells (Koval, 1983a,b) (Fig. 1). The doses necessary to completely inhibit recovery of proliferating *in vitro* populations of these cells correspond to some degree with the *in vivo* doses needed to completely sterilize adult lepidoptera (LaChance *et al.*, 1967). Degree of radiosensitivity appears to be a characteristic that is maintained by other insect orders as well, with cells from dipteran insects (flies and mosquitoes) three to nine times more resistant than mammalian cells, and orthopteran (roaches and grasshoppers), coleopteran (beetles and weevils), and noninsect acarine (ticks and mites) cells intermediate in

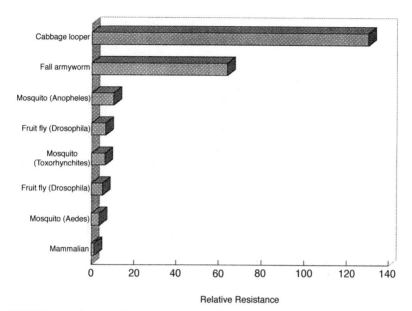

FIGURE 1. Relative radiation resistance of lepidopteran (cabbage looper, *Tricho-plusia ni*; fall armyworm, *Spodoptera frugiperda*), dipteran (mosquito, *Anopheles stephensi* and *Toxorhynchites amboinensis*; fruit fly, *Drosophila melanogaster*; yellow fever mosquito, *Aedes aegypti*), and mammalian (Chinese hamster, *Cricetulus griseus*) cells.

radiosensitivity between dipteran and lepidopteran cells (Koval, 1983a,c). These findings are consistent with the *in vivo* radiosensitivity of these organisms (LaChance and Graham, 1984).

2.2. Ultraviolet Light

Given the natural environment of the parent organisms, it may not be surprising that lepidopteran, cabbage looper, cells are also quite resistant to 254-nm ultraviolet (UV) light. These cells are over 10 times more resistant than mammalian cells, approximately 4 times greater than *Drosophila* cells, several times higher than rat kangaroo, chick embryo, fish, and frog cells, as well as cyanobacteria and several other prokaryotes (Koval, 1986a; Koval et al., 1977).

2.3. Chemical Agents

The sensitivity of cabbage looper cells to a variety of chemical agents including methyl methanesulfonate (MMS), N-methyl-N'-nitro-N-nitrosoguanidine (MNNG), propane sultone (PS), mitomycin C (MMC), and 4-nitroquinoline 1-oxide (4NQO) has been examined (Koval, 1991). Again, the lepidopteran cells were at least 10- to 100-fold more resistant to these agents than mammalian cells. Lepidopteran cells have also displayed resistance to rotenone although the degree of resistance appears to be dependent on the tissue of origin of the cells (Yanagimoto and Mitsuhashi, 1996). Similar resistance has been observed to a variety of other chemicals.

2.4. Heat

Among the insects, lepidopteran cells are much more resistant to heat treatment than dipteran cells (Koval and Suppes, 1992a; Tomasovic and Koval, 1985). Although both cell types are routinely cultured at 28°C, the heating time at 33°C that resulted in 50% cell survival was approximately 30 min for dipteran *Drosophila* cells but slightly more than 48 hr for cabbage looper cells, nearly 100 times longer. Similar relationships existed at higher temperatures as well. At 42°C, *Drosophila* cell survival fell to 50% in about 3 min, whereas at 41.5°C, looper cell survival did not reach 50% until approximately 3 hr. Even at 44°C, the lepidopteran cell survival fell to 50% only after about 15 min.

In fact, the sensitivity of looper cells to heating at 41.5 and 44°C is similar to that of mammalian cells. For instance, survival was reduced to 50% of control after 3 hr of heating at 41.5°C for cabbage looper cells, or after about 1 to 4 hr of heating at the same temperature for various types of mammalian cells. These similarities in thermal sensitivity become remarkable when it is considered that heating to 41.5 and 44°C results in temperature elevations of 13.5 and 16°C for the lepidopteran cells (maintained at 28°C) versus 4.5 and 7°C for mammalian cells (maintained at 37°C). On this basis then, the looper cells exhibit a greatly enhanced relative resistance to heat over mammalian cells.

Although a detailed presentation will not be included in this chapter, it should be noted that lepidopteran cells are also quite resistant to cold temperatures and freezing treatments (Koval, 1996).

3. DNA REPAIR PROCESSES

3.1. X and Gamma Rays

3.1.1. Excision Repair

Unscheduled DNA synthesis (UDS) is a simple method of estimating the amount of DNA excision repair in cells. Using this technique to determine whether x-ray-induced excision repair might explain the radioresistance of cabbage looper cells, it was found that cultured lepidopteran cells performed approximately 20 times the amount of UDS as Chinese hamster cells (Koval *et al.*, 1978). Although this is a sensitive and quantitative assay, it is dependent on several cellular factors such as the amount of DNA damaged, the amount of DNA per cell, the specific activity of cellular thymidine pools, the average size of repaired regions, and the number of repaired sites per unit length of DNA. Experiments to determine which of these factors might explain the high yield of UDS in lepidopteran cells after x-irradiation suggest that they have either a larger average size for x-ray repaired regions or a greater number of repaired regions per unit length of DNA.

3.1.2. Single-Strand Break Repair

DNA single-strand breaks (SSB) are a common type of damage induced by ionizing radiation and their rejoining has been demonstrated in virtually all prokaryotic and eukaryotic systems studied. The inability of cells to repair such damage has been associated with numerous biological dysfunctions. The rate of DNA SSB rejoining after a 100-Gy x-irradiation, as measured by alkaline sucrose gradient sedimentation, was somewhat higher in cabbage looper cells than in hamster cells, with most breaks in the looper cells being repaired within 10 min and in hamster cells within 20 min (Koval *et al.*, 1979). Even though repair of DNA SSB was more rapid in the insect cells, the hamster cells also rejoined the breaks in a relatively short period of time, and considering confounding factors (discussed in Koval *et al.*, 1979), it is difficult to associate the radioresistance of the looper cells with DNA SSB rejoining capabilities.

3.1.3. Double-Strand Break Repair

DNA double-strand breaks (DSB) are believed by many to be the critical molecular lesion leading to the death of cells exposed to ionizing radiation. Hence, the induction and rejoining of DNA DSB were compared in cabbage looper and hamster cells (Koval and Kazmar, 1988) using the technique of neutral elution (Bradley and Kohn, 1979). The initial rates of repair of DSB induced by 90.2 Gy were nearly identical for both cell types, with 50% of the breaks rejoined in approximately 12 min. This indicates that the rate of DSB repair is not a major factor associated with radioresistance. The looper cells reached a final level of about 83% of the DSB rejoined by 1 hr postirradiation. The hamster cells rejoined about 87% of the breaks by 1 hr and continued rejoining to a final level of approximately 92% by 2 hr postirradiation. No further rejoining above these levels occurred in either cell line through at least 6 hr postirradiation. Therefore, neither the rate nor the final extent of DSB rejoining correlates with cell lethality and cannot explain the radioresistance of the lepidopteran cells.

Because maximal rejoining occurred in both cell lines by 3 hr postirradiation, the amount of rejoining at 3 hr was evaluated at various dose levels (Fig. 2). For hamster cells, the percentage of DSB rejoined at 3 hr postirradiation was inversely proportional to dose from 22.6 to 226 Gy. This is consistent with the findings of other groups using hamster and other mammalian cells (Blocher and Pohlit, 1982). The amount of DSB rejoining in cabbage looper cells also declined with increasing gamma-ray doses except the decline was much smaller than for hamster cells, dropping from 86 to 78% for doses of 22.6 to 226 Gy, respectively. Over the same dose range, the amount of DSB rejoining in hamster cells decreased from 96 to 67%. Although this indicates a greater capacity of the looper cells to rejoin DSB at very high doses, it essentially eliminates DNA DSB rejoining capacity as a basis for the marked resistance at lower doses.

The presence of an increased number of small DNA pieces in unirradiated looper cells over unirradiated hamster cells was indicated in DSB experiments (Koval and Kazmar, 1988). This may be

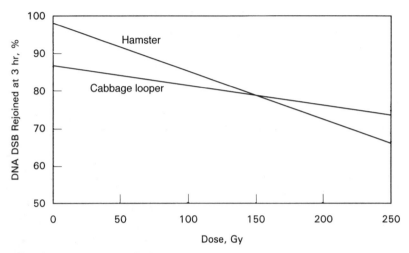

FIGURE 2. Percentage of DNA DSB rejoining in cabbage looper and Chinese hamster cells at 3 hr postirradiation.

related to higher steady-state (background) levels of DSB in the looper cells or may have to do with their much smaller chromosomes (see Section 3.4). Correspondingly, by extrapolating from the data in Fig. 2, it can be estimated that approximately 1 and 14% of the DSB for the hamster and looper cells, respectively, remain at zero dose. This may be evidence that lepidopteran cells can tolerate a higher background level of DSB and function in a relatively normal manner, even with 14% residual DNA DSB.

3.2. Ultraviolet Light

3.2.1. Excision Repair

Given the correlation between the resistance of cabbage looper cells to x rays and the associated large amount of excision repair, it was of interest to perform a similar comparison with UV, especially given the additional resistance of these cells to UV light. UDS was measured in looper and hamster cells as previously described for x rays (Koval *et al.*, 1977). Taking into account the greater amount of DNA per hamster cell, the amount of incorporation was about the same for a given amount of DNA in both cell types. The repair kinetics were similar for both cell types and the repair process was not saturated in the dose range studied. Therefore, excision repair does not explain the greater survival of the lepidopteran cells to UV light.

3.2.2. Photoreactivation

Photoreactivation (PR) is an enzymatic process mediated by exposure to light in the approximate range of 310 to 480 nm; until recently, it was thought to be specific for the repair of cyclobutane-type pyrimidine dimers in DNA. The presence in *Drosophila* of a photoreactivating enzyme specific for the repair of DNA 6-4 photoproducts also has been reported (Todo *et al.*, 1993), thus broadening our view of potential photoreactive capabilities. Pyrimidine dimers represent a major component of the damage induced by 254-nm UV. PR in insect cells was recognized over 40 years ago (Perlitsch and Kelner, 1953). Other processes have since been studied and, more specifically, the destruction of pyrimidine dimers in insect cell DNA (Muraoka *et al.*, 1980) and mRNA (Jackle and Kalthoff, 1980) by photoreactivating illumination has been reported. DNA photoreactivating enzyme has been isolated from silkworm eggs, another lepidopteran insect (Trosko and Wilder, 1973). Cabbage looper cells are quite proficient at repairing UV damage by PR with nearly 70% of the damage being photoreactivable. Depending on the 254-nm UV fluence, survival in lepidopteran cells can be increased as much as 23 times by PR. Although *Drosophila* cells were more sensitive to UV than looper cells, they were capable of photoreactivating nearly 80% of the UV damage, perhaps implying a larger role for cyclobutane-type pyrimidine dimers or 6-4 photoproducts in the lethality of these cells (Koval, 1987). Photoreactivation has been found to reverse the majority of UV-induced blockage of DNA fork progression in lepidopteran cells (Styer and Griffiths, 1992). Although the looper cells are quite adept at photoreactivating DNA damage induced by 254-nm light, it should be remembered that the UV resistance demonstrated, even in the absence of photoreactivating light, was at least 12 times greater than for hamster cells.

3.3. Chemicals

As for x rays and UV, lepidopteran cells were also quite capable of rejoining damaged DNA following treatment with doxorubicin (Bonner *et al.*, 1991) and repairing cross-linking damage induced by 8-methoxypsoralen and 365-nm light (PUVA) (unpublished results). However, these types of repair were not performed more rapidly or to a greater extent than in mammalian cells. It has been reported, though, that mosquito (diptera) cells are more adept at repairing DNA strand breaks than mammalian cells after treatment with bleomycin, an x-ray mimetic drug (Bianchi and Lopez-Larraza, 1991). Therefore, as indicated earlier, lepidopteran and perhaps other insect cells may be superior in repairing x-ray-like DNA damage but about the same as mammalian cells in repairing various other types of DNA damage.

3.4. Chromosome Considerations

There does not appear to be a correlation between chromosome number and radiosensitivity in insects (LaChance *et al.*, 1967). Lepidopteran chromosomes are particularly small and generally thought to be holokinetic (Wolf, 1994; Koval, 1983b). Holokinetic implies that kinetochores or centromeres are spread out or diffused along the chromosome (Suomalainen, 1953; White, 1973). This feature could have a role in radioresistance. Whereas chromosome breakage caused by radiation damage would be expected to result in the loss of chromosome parts and subsequent cell death in species having monokinetic (a single centromere) chromosomes, chromosome fragments would more likely be retained in species having holokinetic chromosomes, therefore reducing the amount of cell killing. Although this is an appealing hypothesis to explain radioresistance in Lepidoptera, both the Hemiptera (true bugs) (Hughes-Schrader and Schrader, 1961) and Homoptera (aphids and leafhoppers) (Schrader, 1947) have holokinetic chromosomes, and yet are sterilized by much lower x-ray doses than the Lepidoptera. Cells of Hemiptera are reported to have nearly twice as much holocentric activity as the Lepidoptera and yet the Hemiptera are much more radiosensitive (Berg and La-Chance, 1976; Gassner and Klemetson, 1974). Another consideration is that lepidopteran cells are unusually resistant to killing by UV light as described earlier. Because chromosomal breakage is not a major consequence following UV exposures, holokinetic chromosomes would not be expected to confer a significant survival benefit to lepidopteran cells in this case. Therefore, holokinetic chromosomes do not appear to be the determinant factor underlying lepidopteran radioresistance even though some role cannot be excluded. Alternatively, the small size and large number of lepidopteran chromosomes may have some influence on radio-resistance. Each of the factors mentioned, large number, small size, and holokinetic nature, are compatible with the earlier speculation that the looper cells are able to remain viable in the presence of a relatively large number of DNA DSB.

3.5. Metabolic Considerations

Insect cells utilize a transamination reaction in their intermediary metabolism whereby glutamic acid and pyruvic acid are converted to α-ketoglutaric acid and alanine (Koval et al., 1976; Grace and Brzostowski, 1966). It has been suggested that amino acid transamination for use in the tricarboxylic acid cycle and its resultant stimulation of respiration favor cell growth and viability in γ-irradiated Escherichia coli and DNA repair in UV-irradiated E. coli (Swenson et al., 1971, 1975). Irradiated cabbage looper cells have a higher rate of transamination (Koval et al., 1976) and a much greater respiration rate, probably because of the uncoupling of oxidative phosphorylation (Koval et al., 1975), than untreated cells. These alterations may be associated with repair processes and the ability of the lepidopteran cells to survive high doses of x-irradiation. Regarding the transaminations, it is worth noting that the hemolymph of insects contains very high concentrations of free amino acids (Wyatt, 1961) in comparison with the blood of mammals (Gerke et al., 1968).

3.6. Enigmatic Relationship between DNA Repair and Stress Resistance

Understanding the precise nature of lepidopteran cell resistance to the various physical and chemical agents is clouded by the apparent inconsistencies in the data obtained (Table I). On one hand, lepidopteran cells demonstrate an exceptional capacity to perform x-ray-induced excision repair and are somewhat better at rejoining radiation induced DNA SSB, which would be expected to greatly enhance the probability of survival. On the other hand, repair of radiation-induced DNA DSB and UV-induced excision repair are not considerably different in magnitude from hamster or other mammalian cells. Although lepidopteran cells are quite

TABLE I
Comparative Repair and Recovery
Processes in Cabbage Looper
and Hamster Cells

Process	Cabbage looper cells	Hamster cells
X-, γ-ray survival	+++	+
X-, γ-ray repair		
Excision	+++	+
SSB	++	+
DSB[a]	+	+
X-, γ-ray recovery		
Liquid holding	++	+
Split dose	++	+
Inducible	++	+
UV survival	+++	+
UV repair		
Excision	+	+
UV recovery		
PR	+++	—[b]
Liquid holding	+	++
Chemical survival	+++	+
Chemical repair		
Cross-linking	+	+
Chemical recovery	varies	varies
Heat survival	+++	+
Heat recovery		
Fractionated	+++	+

[a]Repair kinetics are similar for both cell types; a greater dose dependency on the final amount of rejoining occurs for the hamster cells, whereas the final amount of rejoining in cabbage looper cells decreases only slightly as dose increases.
[b]Not generally observed in hamster cells.

proficient at photoreactivation of damage induced by 254-nm light, they are outperformed to some degree in this regard by dipteran cells, which are vastly greater in sensitivity to 254-nm light. Looper cells are much more resistant to DNA–protein cross-linking treatments or agents than mammalian cells. However, they display about the same amount of cross-linking repair as mammalian cells. Although these findings are not necessarily incompatible, they are difficult to completely reconcile in terms of a direct relationship between DNA repair capability and relative stress resistance. Therefore, although it is reasonable to attribute some portion of the pronounced resistance of lepidopteran cells to DNA repair processes, other mechanisms that are perhaps peripherally associated with DNA repair or are unrelated to DNA repair probably play a prominent role in lepidopteran cell resistance to stresses.

4. INDUCIBLE RECOVERY PROCESSES

4.1. X and Gamma Rays

4.1.1. Multiphasic Survival Curve

The multiphasic nature of the x-ray survival curve for cabbage looper cells offered the first clue indicating a radiation-inducible recovery process (Koval, 1984). Typical cell survival curves (a plot representing the fractional survival of cells versus the amount of stress or treatment delivered) are most commonly composed of a shoulder followed by an exponential straight-line response as dose increases. Although the width of the shoulder and the slope of the exponential portion may vary, the basic shape of the curve usually does not. Subsequent investigations demonstrated that degree of oxygenation, a potential mediator of radiation effects, the existence of a genetically resistant subpopulation of cells, and cell cycle variations in radiosensitivity could not explain the observed radioresistance or the multiphasic curve shape.

4.1.2. Split-Dose and Delayed Plating Experiments

As another means of revealing an inducible recovery process in the looper cells, experiments involving split doses and delayed plating following irradiation were performed (Koval, 1988). For delayed plating experiments, looper cells entering stationary growth phase were irradiated with gamma rays and then held for various periods of time before cell dilution and plating for colony formation. As observed for mammalian and other cells, the surviving fraction increased severalfold over cells diluted and plated immediately after irradiation. For split dose experiments, the cells were incubated for various periods of time between two equivalent

doses of radiation and then plated for colony formation immediately following the second dose. Again, in a manner similar to
mammalian cells, survival was considerably greater in looper cells
treated in this manner over cells plated immediately following a
single dose equal to the sum of the split doses. Both delayed
plating and split-dose recovery processes exhibited similar biphasic recovery kinetics and reached maximal levels by 6 hr.
The survival levels attained by cells having undergone these processes are quite high compared with mammalian cells and the
similar kinetics of both processes suggest a common underlying
mechanism.

Although mammalian cells also undergo similar recovery processes, these studies demonstrate that cabbage looper cells exhibit a type of recovery that leads to unexpectedly high levels of
survival beyond those previously described. The split-dose experiments especially present further evidence of the existence of an
inducible recovery process. For example, survival of cells irradiated with a single dose of 150 Gy was near 15%. The survival of
cells receiving 153 Gy followed 6 hr later by another 153 Gy, a
total treatment dose of 306 Gy, was approximately 77%. If the
damage produced by the first of the 153-Gy splits was completely
repaired, one might optimistically expect the survival to be dependent on the second dose alone, or near 15%. Instead, the survival
was approximately five times greater, indicating not only that the
damage induced by the first dose had been repaired to a large
extent, but also that the damage induced by the second dose was
repaired to a greater extent than if the cells had received only one
of the doses.

The delayed plating and split-dose values that represent maximal recovery (6 hr data) decreased very slightly as total dose
increased up to about 300 Gy but then dropped sharply around
400 Gy (Fig. 3). This steep drop in survival paralleled a similar
precipitous drop in the same dose range of the multiphasic curve
obtained by irradiating and immediately plating the moth cells.
This is consistent with interpreting the broad inflection region on
the cell survival curve as resulting from an inducible process and
the drop in survival between 300 and 400 Gy as saturation of this
process. The data therefore suggest that the maximal amount of
delayed plating or split-dose recovery represents the maximal
amount of repair that can be induced in cells. This further indi-

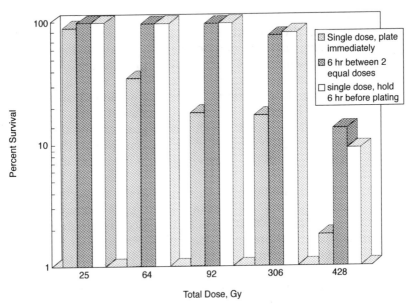

FIGURE 3. Survival differences between irradiating cabbage looper cells with a single dose and plating immediately versus delaying plating for 6 hr or allowing 6 hr between two equivalent doses.

cates that the highest dose on the inflection portion of a multi-phasic survival curve corresponds to the saturation point of the inducible repair process and implies an association between the inducible response and delayed plating and split-dose recovery.

4.1.3. Induction of RNA and Protein Synthesis

Investigations were performed to determine whether the im-plied inducible response involved *de novo* RNA and protein syn-thesis. RNA and protein synthesis were inhibited in cabbage looper cells during split dose and delayed plating recovery condi-tions (Koval, 1986b). When the cells were incubated in cyclohex-imide or actinomycin D following irradiation, but prior to cell dilution and plating, no increase in survival was observed (Fig. 4). Correspondingly, when the cells were incubated in cycloheximide

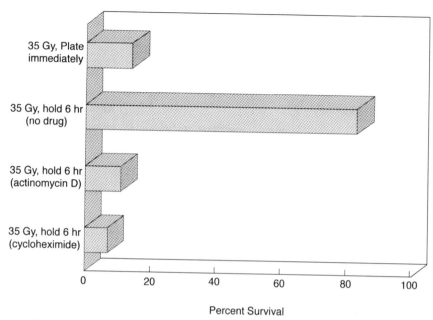

FIGURE 4. Survival comparison between cabbage looper cells plated immediately after a single irradiation and those held in medium with or without actinomycin D or cycloheximide for 6 hr before plating.

during the interval between split radiation doses, no increase in survival occurred. The concentrations of cycloheximide and actinomycin D that were used in these studies inhibited protein and RNA synthesis, respectively, and caused little or no toxicity to control cells. In addition, when cycloheximide was added prior to the first irradiation, but removed during the interval between split doses, survival increased as for cells treated with radiation only, indicating that the lack of survival increase for the reciprocal treatment was not related to synergistic effects between the radiation and cycloheximide. The absence of an increase in survival in the presence of cycloheximide and actinomycin D demonstrates

that protein and RNA synthesis are essential for development of
the enhanced survival.

4.1.4. Protein Identification

Subsequent to determining the importance of induced protein
synthesis, experiments were performed to begin to identify and
characterize the radiation-inducible proteins in cabbage looper
cells (Rand and Koval, 1994). Such proteins were sought in both
nuclear and cytoplasmic fractions of irradiated cells as well as
RNA from these cells which was translated *in vitro*. The cells were
treated with 25, 64, and 350 Gy of γ rays resulting in 90, 35, and
6% survival, respectively. Two-dimensional polyacrylamide gel
electrophoresis revealed the *de novo* synthesis of several proteins
as well as the complete inhibition of others. Nineteen proteins were
identified and ranged in size from 18 to 73 kDa, with a *p*I distribution of 4.7 to 6.1. In addition to the unique proteins, a large number of other proteins were also either up- or downregulated. Similar observations were made across the dose range studied in both
nuclear and cytoplasmic fractions as well as in the translation
products of RNA produced after irradiation. This coordinated regulation of protein synthesis appears to be associated with the
pronounced resistance to stress displayed by these cells. The
function(s) of the inducible proteins are unknown.

4.2. Ultraviolet Light

4.2.1. Biphasic Survival Curve

Evidence indicating that an inducible recovery process is responsible for the multiphasic x-ray survival response for cabbage
looper cells was presented in Section 4.1.1 (Koval, 1984). Similarly,
a biphasic response was obtained for 254-nm UV and present
speculation is that the looper cells also possess an inducible recovery process that is associated with this survival response, as well.

4.2.2. Delayed Plating Experiments

Holding cells before plating is responsible for as much as a
2-fold increase in cabbage looper cell survival following UV exposure (Koval, 1986b). This is a rather small increase when com-

pared with the 23-fold survival enhancement resulting from PR as discussed in Section 3.2.2. This suggests that the primary lethal lesions induced by UV light are the pyrimidine dimer and/or the 6-4 photoproduct and that PR is the preferred means of repairing these lesions in looper cells. This is supported by the relatively small amount of UV-induced UDS (Koval *et al.*, 1977), a process that may reflect the molecular repair mechanism responsible for delayed plating recovery, i.e., excision repair (Konze-Thomas *et al.*, 1979; Weichselbaum *et al.*, 1978).

4.3. Chemicals

4.3.1. Split-Dose and Delayed Plating Studies

Concentrations of MMS, MNNG, PS, MMC, and 4NQO that reduced cell survival to about 10% were used to assess recovery ability assayed by colony formation in split-dose and delayed plating experiments (Koval, 1991). Looper cells were able to recover from MMS, MNNG, and PS in both types of experiments. Recovery from 4NQO was observed in delayed plating experiments and not assessed in split-dose experiments. In all cases where recovery was observed, survival enhancement was approximately twofold. Recovery from MMC (a cross-linking agent) exposure was not observed in either type of experiment. In addition, recovery from 8-methoxypsoralen or angelicin plus UVA light (PUVA), other cross-linking treatments, was not observed (Koval and Suppes, 1992b). Therefore, even though the cells are quite resistant to killing by these cross-linking agents, the type of damage produced may not stimulate or may otherwise preclude the induction of a recovery process.

4.4. Heat

DNA is believed to be the major target for killing by each of the agents discussed in this chapter except for heat. Therefore, as the molecular alteration that initiates the response is likely to be different than for the other agents, examining studies with heat in conjunction with studies on the other agents has the potential to contribute significantly toward understanding recovery process induction mechanisms.

4.4.1. Fractionated Heating

Cellular Recovery. Thermotolerance, a classical induced response, was demonstrated in cabbage looper cells at 33, 37, 41.5, and 44°C (Koval and Suppes, 1992b) (Fig. 5). This is in contrast to the slight thermotolerance observed in *Drosophila* cells at 33°C, but not at 37 or 42°C (Tomasovic and Koval, 1985). Thermotolerance was assessed by fractionated heating experiments for both cell types. The thermotolerance induction conditions used for *Drosophila* cells were more damaging than those used for the looper cells. The corresponding survival levels were about 0.5 to 0.01 for the *Drosophila* cells and 1.0 to 0.2 for the looper cells. Therefore, higher tolerance levels are not correlated with more heat killing by the priming heat dose as might be expected. Although it remains to be seen whether these induction conditions would make a difference in development of thermotolerance, it is clear that several distinct differences exist regarding the survival response to heat in these two cultured insect cell systems. Additionally, the looper cells develop a significant amount of thermotolerance in fractionated heating experiments at 44°C in a manner quite comparable to several types of mammalian cells (Gerweck and DeLaney, 1984; Nielsen and Overgaard, 1982; Henle and Dethlefsen, 1978).

Molecular Recovery. The patterns of heat-induced inhibition of overall DNA and protein synthesis are generally similar to those that have been reported for various mammalian and human cells (Koval and Suppes, 1992b; Oesterreich *et al.*,1990; Kern *et al.*, 1988; Warters and Stone, 1983; Henle and Leeper, 1979). Inhibition of both types of synthesis increased with time and temperature. Recovery was somewhat faster for protein synthesis than DNA synthesis, again in line with the previous studies (Warters and Stone, 1983; Henle and Leeper, 1979). When a priming heat exposure was used to induce thermotolerance prior to a second heat treatment, DNA synthesis was reduced to a level as low as or below the level that would have been produced by the second treatment alone. However, after a 41.5°C priming dose, protein synthesis was less inhibited and its recovery was more rapid in tolerant cells than in cells receiving only a single 44°C treatment. For 37 and 44°C priming regimens, the tolerant cells displayed

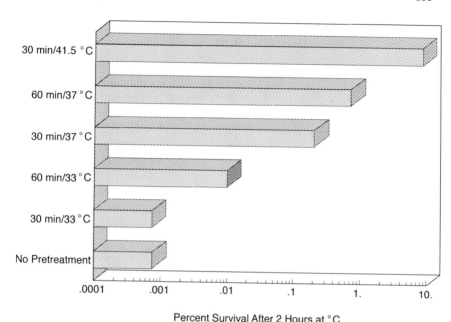

FIGURE 5. Survival of cabbage looper cells after a treatment regimen beginning with the conditions shown, followed by 2 hr of incubation at 28°C, and then heating for 2 hr at 44°C.

similar protein synthesis kinetics to the nontolerant cells. The degree of inhibition in the tolerant cells was therefore dependent on the time and temperature of the priming heat exposure. This is relatively consistent with other reports (Lee *et al.*, 1990; Oesterreich *et al.*, 1990), although higher tolerance levels did not correlate with the amount of heat killing by the priming dose. Investigations with other eukaryotic cell systems have also demonstrated that alterations in protein synthesis are not necessary for the expression of thermotolerance (Watson *et al.*, 1984; Tomasovic *et al.*, 1983). Therefore, despite the marked heat resistance observed in the looper cells, fundamental alterations in DNA and protein synthesis appear to be comparable to mammalian and human cells. The teleological significance of a rapid and prolonged depression in DNA synthesis (to allow time for cellular repair and recovery) and the perhaps more intricate balance in protein synthesis inhibition and recovery have been discussed at length elsewhere (Lee *et al.*, 1990; Lepock *et al.*, 1990; Wong and Dewey, 1982; Warters, 1988).

5. GENETIC BASIS FOR CELLULAR RADIORESISTANCE OF LEPIDOPTERA

DNA from cabbage looper cells was introduced into (1) Chinese hamster cells that exhibit typical mammalian cell radioresistance and (2) a hypersensitive mutant of the hamster cells, to determine whether the radiation sensitivity of these mammalian cells could be altered by lepidopteran genes (Drabek *et al.*, 1993). The looper cell DNA was cotransfected into the mammalian cells with a plasmid containing an antibiotic resistance gene. Cells resistant to the antibiotic gentamycin were then selected to identify transformation-competent cells. Cells selected by this screening as well as the unaltered mammalian lines and the mammalian cells containing only the antibiotic-resistance gene but not the lepidopteran DNA, were irradiated at low dose rates with γ-ray doses known to be lethal to the respective mammalian lines. The *only* cells surviving these treatments were the hypersensitive, but not the normal wild-type, hamster cells that had been transfected with both the looper cell DNA and the antibiotic-resistance gene. Fourteen such clones of cells were isolated. DNA from these clones was subjected to Southern analysis with DNA probes specific to lepidopteran DNA (a transposon plasmid and arylphorin, a chitin protein, cDNA). This analysis confirmed the presence of lepidopteran DNA in the clones. Three of those clones having the greatest amount of homology with the lepidopteran probes were subjected to more in-depth assessment of their radiosensitivity. This assessment indicated that the radiation sensitivity of the originally hypersensitive hamster cell mutant had been restored to nearly that of the normal hamster cell line.

Given the large number of genes likely to be involved in a cell's response to radiation injury, it is difficult to conceive of restoring mutant sensitivity to a level above that of the wild-type sensitivity.

For example, the genetic basis of a *defect* can be corrected, but cannot surpass the capacity of the original system. By analogy, the heart of a young healthy Olympic athlete could be transplanted into an elderly sedate individual; however, it would be unreasonable to then expect the elderly person to perform in the decathlon. The outcome of the transfection experiment, then, is probably the best that one might expect. Further, this may explain why no transfected wild-type hamster cells were isolated during the radiation screening procedure described above (i.e., the wild-type system was already functioning at its maximum capacity). These experiments provide evidence for a genetic basis for radiation resistance and provide a resource for identifying the genes responsible for this trait.

6. SUMMARY

6.1. Conclusions

Cells cultured from insects are significantly more resistant to killing by a number of physical and chemical agents than cultured mammalian cells. The degree of radioresistance corresponds to insect order, with lepidoptera being by far the most resistant. The role of DNA repair processes in the radioresistance of lepidopteran cells remains to some degree an enigma. It does appear certain, however, that an inducible recovery mechanism is at least partially responsible for the magnitude of lepidopteran cell resistance. This inducible system is dependent on transcriptional and translational activities that are not present in unirradiated cells or cells receiving less than some minimal amount of radiation (or other stress) necessary for activation of the process. The result of such activity is the *de novo* production of several new proteins, the complete inhibition of others, and wide-ranging levels of up- and downregulation of a large number of other proteins. A function has not been determined for any of these proteins. DNA transfection studies suggest that the genetic information responsible for radiation resistance can be transferred and confer resistance to sensitive cells of other eukaryotic species. Inducible mechanisms have also been associated with resistance of the lepidopteran cells to 254-nm UV light and heat exposures.

6.2. Advantages of Using Lepidopteran Cells

Perhaps the major advantage of the lepidopteran cell line model is that using mammalian or most other higher eukaryotic cells results in very few survivors at the doses of various stresses required for accurate studies of molecular lesions. This results in nonsurviving cells being the objects of study when attempting to dissect DNA repair or other subcellular mechanisms. The lepidopteran cells exhibit such a pronounced radiation resistance that large numbers of surviving cells are present even at the relatively high doses necessary to evaluate molecular damage. Hence, surviving lepidopteran cells might provide more relevant general information regarding mechanisms that are valuable in a cell's ability to survive various stress exposures than nonsurviving cells of mammalian or other species. Another advantage of using the lepidopteran cells, which is a consequence of their resistance, is the broad dose range used in survival studies with various stressful agents. The broad dose range makes the system more amenable than mammalian or other cells in detecting subtleties dealing with the cell survival response.

6.3. Implications

Lepidopteran cells have the potential to provide information that will contribute to a better understanding of cell and molecular mechanisms of stress resistance. The specific molecular basis for the profound resistance of lepidopteran cells may have far-reaching implications from an evolutionary perspective as well as having an impact on our fundamental understanding of the complex mechanisms involved in the stress response in eukaryotic cells. Understanding the lepidopteran stress resistance process could provide the basis for intervention to obstruct the process, thereby having great practical relevance to such diverse fields as insect pest control and the radiotherapy of human tumors.

REFERENCES

Berg, G. J., and LaChance, L. E., 1976, Dominant lethal mutations in insects with holokinetic chromosomes: Irradiation of pink bollworm sperm, *Ann. Entomol. Soc. Am.* **69:**971–976.

Bianchi, N. O., and Lopez-Larraza, D. M., 1991, DNA damage and repair induced by bleomycin in mammalian and insect cells, *Environ. Mol. Mutagen.* **17:**63–68.

Blocher, D., and Pohlit, W., 1982, DNA double strand breaks in Ehrlich ascites tumour cells at low doses of X-rays. II. Can cell death be attributed to double strand breaks? *Int. J. Radiat. Biol.* **42**:329–338.

Bond, V. P., Fliedner, T. M., and Archambeau, J. O., 1965, *Mammalian Radiation Lethality*, Academic Press, New York.

Bonner, J. A., Christianson, T. H., and Koval, T. M., 1991, Correlation of doxorubicin sensitivity with the stabilization of DNA topoisomerase II complexes in an extremely doxorubicin-resistant lepidopteran insect cell line, *Proc. Am. Assoc. Cancer Res.* **32**:338.

Bradley, M. O., and Kohn, K. W., 1979, X-ray induced DNA double strand break production and repair in mammalian cells as measured by neutral filter elution, *Nucleic Acids Res.* **7**:793–804.

Casarett, A. P., 1968, *Radiation Biology*, Prentice–Hall, Englewood Cliffs, NJ.

Drabek, R., Koval, T. M., Stamato, T., Vannais, D., and Waldren, C., 1993, Transfection of CHO and CHOXR1 with DNA from TN-368 lepidopteran cells: Selection for transformants hyperresistant to γ-rays, *Abstracts of the 41st Annual Meeting of the Radiation Research Society*, p. 139.

Ducoff, H. S., 1972, Causes of death in irradiated adult insects, *Biol. Rev.* **47**: 211–240.

Gassner, G., and Klemetson, D. J., 1974, A transmission electron microscope examination of hemipteran and lepidopteran gonial centromeres, *Can. J. Genet. Cytol.* **16**:457–464.

Gerke, C. W., Zumwalt, R. W., Stalling, D. L., and Wall, L. L., 1968, Quantitative gas–liquid chromatography of amino acids in proteins and biological substances, Analytical Biochemistry Laboratories, Inc., Columbia, MO.

Gerweck, L. E., and DeLaney, T. F., 1984, Persistence of thermotolerance in slowly proliferating plateau–phase cells, *Radiat. Res.* **97**:365–372.

Grace, T. D. C., and Brzostowski, H. W., 1966, Analysis of the amino acids and sugars in an insect cell culture medium during cell growth, *J. Insect Physiol.* **12**:625–633.

Henle, K. J., and Dethlefsen, L. A., 1978, Heat fractionation and thermotolerance: A review, *Cancer Res.* **38**:1843–1851.

Henle, K. J., and Leeper, D. B., 1979, Effects of hyperthermia on macromolecular synthesis in Chinese hamster ovary cells, *Cancer Res.* **39**:2665–2674.

Hughes-Schrader, S., and Schrader, F., 1961, The kinetochore of the Hemiptera, *Chromosoma* **12**:327–350.

Jackle, H., and Kalthoff, K., 1980, Photoreversible UV-inactivation of messenger RNA in an insect embryo (*Smittia* spec. Chironomidae, Diptera), *Photochem. Photobiol.* **32**:749–761.

Kern, D. H., Krag, D. N., Kauffman, G. L., Morton, D. L., and Storm, F. K., 1988, Thermal resistance of human malignant melanoma modulated by prostaglandin E_2, *J. Surg. Oncology* **37**:60–64.

Konze-Thomas, B., Levinson, J. W., Maher, V. M., and McCormick, J. J., 1979, Correlation among the rates of dimer excision, DNA repair replication, and recovery of human cells from potentially lethal damage induced by ultraviolet radiation, *Biophys. J.* **28**:315–326.

Koval, T. M., 1983a, Intrinsic resistance to the lethal effect of X-irradiation in insect and arachnid cells, *Proc. Natl. Acad. Sci. USA* **80**:4752–4755.

Koval, T. M., 1983b, Radiosensitivity of cultured insect cells: I. Lepidoptera, *Radiat. Res.* **96**:118–126.

Koval, T. M., 1983c, Radiosensitivity of cultured insect cells: II. Diptera, *Radiat Res.* **96**:127–134.

Koval, T. M., 1984, Multiphasic survival response of a radioresistant lepidopteran cell line, *Radiat. Res.* **98**:642–648.

Koval, T. M., 1986a, Enhanced survival by photoreactivation and liquid-holding following UV damage of TN-368 insect cells, *Mutat. Res.* **166**:149–156.

Koval, T. M., 1986b, Inducible repair of ionizing radiation damage in higher eukaryotic cells, *Mutat. Res.* **173**:291–293.

Koval, T. M., 1987, Photoreactivation of UV damage in cultured *Drosophila* cells, *Experientia* **43**:445–446.

Koval, T. M., 1988, Enhanced recovery from ionizing radiation damage in a lepidopteran insect cell line, *Radiat. Res.* **115**:413–420.

Koval, T. M., 1991, Recovery from exposure to DNA-damaging chemicals in radiation-resistant insect cells, *Mutat. Res.* **262**:219–225.

Koval, T. M., 1996, Cold hardiness of cultured lepidopteran cells, *In Vitro Cell. Dev. Biol.* **32**:37A.

Koval, T. M., and Kazmar, E. R., 1988, DNA double-strand break repair in eukaryotic cell lines having radically different radiosensitivities, *Radiat. Res.* **113**: 268–277.

Koval, T. M., and Suppes, D., 1992a, Heat resistance and thermotolerance in a radiation-resistant cell line, *Int. J. Radiat. Biol.* **61**:425–431.

Koval, T. M., and Suppes, D., 1992b, Survival response of TN-368 lepidopteran cells to psoralins and UVA light, *In Vitro Cell. Dev. Biol.* **28**:88a.

Koval, T. M., Myser, W. C., and Hink, W. F., 1975, Effects of X-irradiation on cell division, oxygen consumption, and growth medium pH of an insect cell line cultured *in vitro*, *Radiat. Res.* **64**:524–532.

Koval, T. M., Myser, W. C., and Hink, W. F., 1976, The effect of x-irradiation on amino acid utilization in cultured insect cells, *Radiat. Res.* **67**:305–313.

Koval, T. M., Hart, R. W., Myser, W. C., and Hink, W. F., 1977, A comparison of survival and repair of UV-induced DNA damage in cultured insect versus mammalian cells, *Genetics* **87**:513–518.

Koval, T. M., Myser, W. C., Hart, R. W., and Hink, W. F., 1978, Comparison of survival and unscheduled DNA synthesis between an insect and a mammalian cell line following X-ray treatments, *Mutat. Res.* **49**:431–435.

Koval, T. M., Hart, R. W., Myser, W. C., and Hink, W. F., 1979, DNA single-strand break repair in cultured insect and mammalian cells after X-irradiation, *Int. J. Radiat. Biol.* **35**:183–188.

Labandeira, C. C., and Sepkoski, J. J., 1993, Insect diversity in the fossil record, *Science* **261**:310–315.

LaChance, L. E., and Graham, C. K., 1984, Insect radiosensitivity: Dose curves and dose-fractionation studies of dominant lethal mutations in the mature sperm of 4 insect species, *Mutat. Res.* **127**:49–59.

LaChance, L. E., Schmidt, C. H., and Bushland, R. C., 1967, Radiation-induced sterilization, in: *Pest Control: Biological, Physical, and Selected Chemical Methods* (W. W. Kilgore and R. L. Doutt, eds.), Academic Press, New York, pp. 147–196.

Lee, Y. J., Perlaky, L., Dewey, W. C., Armour, E. P., and Corry, P. M., 1990, Differences in thermotolerance induced by heat or sodium arsenite: Cell killing and inhibition of protein synthesis, *Radiat. Res.* **121**:295–303.

Lepock, J. R., Frey, H. E., Heynen, M. P., Nishio, J., Waters, B., Ritchie, K. P., and Kruuv, J., 1990, Increased thermostability of thermotolerant CHL V79 cells as determined by differential scanning calorimetry, *J. Cell. Physiol.* **142:** 628–634.

Muraoka, N., Okuda, A., and Ikenaga, M., 1980, DNA photoreactivating enzyme from silkworm, *Photochem. Photobiol.* **32:**193–197.

Nielsen, O. S., and Overgaard, J., 1982, Influence of time and temperature on the kinetics of thermotolerance in L_1A_2 cells in vitro, *Cancer Res.* **42:**4190–4196.

O'Brien, R. D., and Wolfe, L. S., 1964, *Radiation, Radioactivity and Insects*, Academic Press, New York.

Oesterreich, S., Benndorf, R. and Bielka, H., 1990, The expression of the growth–related 25kDa protein (p25) of Ehrlich ascites tumor cells is increased by hypothermic treatment (heat shock), *Biomed. Biochim. Acta* **49:**219–226.

Perlitsch, M., and Kelner, A., 1953, The reduction by reactivating light of the frequency of phenocopies induced by ultraviolet light in *Drosophila melanogaster*, *Science* **118:**165–166.

Rand, A., and Koval, T. M., 1994, Coordinate regulation of proteins associated with radiation resistance in cultured insect cells, *Radiat. Res.* **138:**S13–S16.

Schrader, F., 1947, The role of the kinetochore in the chromosomal evolution of the Heteroptera and Homoptera, *Evolution* **1:**134–142.

Styer, S. C., and Griffiths, T. D., 1992, Effect of UVC light on growth, incorporation of thymidine, and DNA chain elongation in cells derived from the Indian meal moth and the cabbage looper, *Radiat. Res.* **130:**72–78.

Suomalainen, E., 1953, The kinetochore and the bivalent structure in the Lepidoptera, *Hereditas* **39:**88–96.

Swenson, P. A., Schenley, R. L., and Boyle, J. M., 1971, Interference with respiratory control by ionizing radiations in *Escherichia coli* B/r, *Int. J. Radiat. Biol.* **20:**223.

Swenson, P. A., Ives, J. E., and Schenley, R. L., 1975, Photoprotection of *E. coli* B/r: Respiration, growth, macromolecular synthesis and repair of DNA, *Photochem. Photobiol.* **21:**235–241.

Todo, T., Takemori, H., Ryo, H., Ihara, M., Matsunaga, T., Nikaido, O., Soto, K., and Nomura, T., 1993, A new photoreactivating enzyme that specifically repairs ultraviolet light-induced (6-4) photoproducts, *Nature* **361:**371–374.

Tomasovic, S. P., and Koval, T. M., 1985, Relationship between cell survival and heat-stress protein synthesis in a *Drosophila* cell line, *Int. J. Radiat. Biol.* **48:** 635–650.

Tomasovic, S. P., Steck, P. A., and Heitzman, D., 1983, Heat-stress proteins and thermal resistance in rat mammary tumor cells, *Radiat. Res.* **95:**399–413.

Trosko, J. E., and Wilder, K., 1973, Repair of UV-induced pyrimidine dimers in *Drosophila melanogaster* cells *in vitro*, *Genetics* **73:**297–302.

Warters, R. L., 1988, Hyperthermia blocks DNA processing at the nuclear matrix, *Radiat. Res.* **115:**258–272.

Warters, R. L., and Stone, O. L., 1983, Macromolecule synthesis in HeLa cells after thermal shock, *Radiat. Res.* **96:**646–652.

Watson, K., Dunlop, G., and Cavicchioli, R. R., 1984, Mitochondrial and cytoplasmic protein syntheses are not required for heat shock acquisition of ethanol and thermotolerance in yeast, *FEBS Lett.* **172:**299–302.

Weichselbaum, R. R., Nove, J., and Little, J. B., 1978, Deficient recovery from potentially lethal radiation damage in ataxia telangiectasia and xeroderma pigmentosum, *Nature* **271**:261–262.

White, M. J. D., 1973, *Animal Cytology and Evolution*, Cambridge University Press, London.

Wolf, K. W., 1994, The unique structure of lepidopteran spindles, *Int. Rev. Cytol.* **152**:1–48.

Wong, R. S. L., and Dewey, W. C., 1982, Molecular studies on the hyperthermic inhibition of DNA synthesis in Chinese hamster ovary cells, *Radiat. Res.* **92**: 370–395.

Wyatt, G. R., 1961, The biochemistry of insect hemolymph, *Annu. Rev. Entomol.* **6**:75–102.

Yanagimoto, Y., and Mitsuhashi, J., 1996, Production of rotenone-inactivating substance(s) by rotenone-resistant insect cell line, *In Vitro Cell. Dev. Biol. Anim.* **32**:399–402.

5

Apoptosis as a Stress Response
Lessons from an Insect Virus

ROLLIE J. CLEM

1. INTRODUCTION: THE DICHOTOMY OF CELL DEATH

Cells have evolved a number of protective mechanisms to help them withstand stressful external stimuli such as heat, toxic chemicals, anoxia, and infection by microorganisms. Despite these protective responses, if the level of stress is too great, cell death results. Kerr *et al.* (1972) were the first to emphasize the differences between cell death directed by the cell, or *programmed cell death*, and cell death beyond the control of the cell, or *necrotic cell death*. Whether the death of a cell is programmed or necrotic often depends on the severity of the stress signal. Cells have evolved the ability to respond to many physiological stimuli by assisting in their own death. However, if the level of stress is so great that it overwhelms the ability of the cell to function, necrosis is often the result.

ROLLIE J. CLEM • Department of Molecular Microbiology and Immunology, The Johns Hopkins School of Hygiene and Public Health, Baltimore, Maryland 21205.

Stress-Inducible Processes in Higher Eukaryotic Cells, edited by Koval. Plenum Press, New York, 1997.

1.1. Necrosis versus Programmed Cell Death

Necrosis and programmed cell death can be distinguished on several levels. Necrosis is the result of cell injury, and is characterized by cell swelling and loss of membrane integrity (Walker *et al.*, 1988). Necrosis is uncontrolled cell death; membranes rupture, leakage of organelles results in loss of cellular integrity and function, and digestive enzymes are released into inappropriate intracellular compartments as well as into the extracellular milieu, resulting in inflammation and tissue injury. The chromatin of necrotic cells is randomly digested by proteases and nucleases, resulting in a smear when visualized by gel electrophoresis. As necrosis is beyond the control of the cell, it is not considered a specific response to the injurious stimulus.

In contrast, programmed cell death infers that the cell actively participates in its own demise via a set of genetic pathways. The ability of cells to actively commit suicide appears to be widely conserved in evolution, and was probably an important step in the rise of multicellular organisms. Apoptosis is a morphologically and biochemically defined type of programmed cell death (Kerr *et al.*, 1972) that has received considerable attention in recent years. Apoptosis appears to be the predominant form of programmed cell death in higher vertebrates. Cell death very similar to vertebrate apoptosis is seen in certain situations in invertebrates although there appears to be more variety in the morphological forms of programmed cell death in these lower animals (Schwartz *et al.*, 1993). From a genetic point of view, because apoptosis is by definition a morphological term, the distinction between apoptosis and programmed cell death is becoming less important as it is becoming clear that the essential genetic pathways for controlling cell death are widely conserved in both lower and higher animals.

The morphology of apoptosis described by Kerr *et al.* is characterized by the formation of large protuberances at the surface of the cell; these pinch off, resulting in the release of membrane-bound bodies (called apoptotic bodies) containing cytoplasmic contents including intact mitochondria and other organelles. During this process of blebbing, the cell appears to shrink (thus the early term *shrinkage necrosis*). Within the nucleus, the chromatin dramatically condenses and moves to the periphery of the nucleus. Later the nucleus also blebs and nuclear contents are released into apoptotic bodies. In tissues, the apoptotic bodies are rapidly phagocytized by neighboring cells. An important distinction between apoptosis and necrosis is that the membrane of a cell dying by apoptosis remains intact both during and after death, until the apoptotic bodies are taken up by neighboring cells. Thus, the contents of the cell are not exposed to the extracellular environment.

As one would expect, these striking morphological changes that occur in cells dying by apoptosis are driven by biochemical changes within the cell. These biochemical reactions leading to or resulting from apoptotic cell death are currently the subjects of intense investigation. One of the first to be described (which thus became a "hallmark" of apoptosis) is that early in the death program, prior to the manifestation of the morphological changes associated with apoptosis, the chromatin is digested into oligonucleosomal length fragments by an endogenous endonuclease. On agarose gel electrophoresis, a ladder of fragments can be seen whose lengths are multiples of 180 bp, which is the distance between nucleosomes. Although this ladder is not always seen in mammalian cells dying by apoptosis, it is quite common. Recently, significant progress has been made in defining other biochemical events associated with apoptosis. A hypothesis is emerging in which any one of multiple signaling pathways is activated by a particular death stimulus, with each of these signaling pathways leading to a common effector, or executioner, pathway. These pathways are beginning to be defined, and have been the subject of many recent reviews (Vaux and Strasser, 1996; Steller, 1995; Thompson, 1995; Oltvai and Korsmeyer, 1994; Vaux et al., 1994; Raff et al., 1993; Williams and Smith, 1993). We now know the identity of a relatively large number of the players involved in these pathways, although in most cases their exact modes of action are still unknown.

1.2. Apoptosis as a Stress Response

In many situations where an overwhelming stress stimulus leads to necrotic cell death, it has been shown that an intermediate level of stress will instead induce apoptotic death (Kerr and Harmon, 1991). The fact that a cell chooses to commit suicide when it is exposed to a high, but not unmanageable level of stress implies that this represents an evolutionary advantage to the organism. It has been suggested that a reason for this seemingly altruistic behavior on the part of a single cell level may be to protect the overall genome integrity of the organism (Kerr and Harmon, 1991). When cells are exposed to stress stimuli, cellular components, including DNA, are damaged. If the level of DNA damage is too high to be successfully repaired before the next cell division, it would be preferable to eliminate the cell rather than pass on mutations that could lead to tumorigenesis. This may be especially true in cells that divide rapidly, such as lymphocytes or fibroblasts. Thus, apoptosis could be considered the ultimate stress response, involving a specific response on the part of the cell so as to avoid a negative outcome for the entire organism.

Another situation where apoptosis may be an advantage for the organism as a whole is during invasion by microorganisms. It has been noted that certain characteristics of apoptosis, including the maintenance of integrity of the plasma membrane throughout the process and the activation of intracellular endonucleases, could act to contain and inactivate infectious organisms such as viruses (Vaux and Hacker, 1995; Martz and Howell, 1989; Clouston and Kerr, 1985). In fact, it appears that the killing of virus-infected cells by cytotoxic T lymphocytes in vertebrates is mediated by apoptosis (Vaux *et al.*, 1994). Thus, an individual infected cell may choose (or be persuaded) to commit suicide to limit spread of the virus. In recent years, a large number of viruses have been shown to induce apoptosis, suggesting that apoptosis may be an important cellular defense against virus infection. Not to be outdone, many viruses, particularly the large DNA viruses, encode proteins that can preclude the apoptotic response, thereby giving them a replicative advantage.

2. THE BACULOVIRUSES

2.1. General Characteristics

The baculoviruses comprise a family of viruses that infect only arthropods [in the interest of brevity, refer to the following recent reviews for most of the background references in this section: Kool *et al.* (1995), Kool and Vlak (1993), O'Reilly *et al.* (1992), Rohrmann (1992)]. Over 400 baculoviruses have been described, most of which were isolated from insects of the order Lepidoptera, although only a small number of these viruses have been characterized at the molecular level. The genomes of baculoviruses consist of double-stranded, circular DNA in the size range of 80 to 250 kbp. Because the DNA is infectious when introduced into cells by transfection, it has been relatively easy to genetically manipulate the genomes of baculoviruses. This ease of manipulation has helped to make the baculoviruses more attractive to study. They have been utilized for two main purposes: (1) as an individual baculovirus usually infects only a small number of related insect species, members of the baculovirus family have been used as highly selective biological control agents (Miller, 1995), and (2) because of the ease with which foreign DNA can be inserted into the baculovirus genome, their ability to replicate in eukaryotic cells (with all of the appropriate eukaryotic posttranslational modification machinery), and the high levels of gene expression that can be obtained, the baculoviruses have also been extensively utilized as vectors for the expression of foreign genes for nearly 15 years (O'Reilly *et al.*, 1992).

The best-studied baculovirus at the molecular level is *Autographa californica* nuclear polyhedrosis virus (AcMNPV; the M is a taxonomic designation that refers to the number of nucleocapsids per virus particle). The entire genome of AcMNPV has been sequenced and is 133,894 bp in length (Ayres *et al.*, 1994). This virus encodes 154 potential open reading frames, of which approximately 50 have been functionally characterized, including a number of genes that have homology to known cellular genes such as homologues of DNA polymerase, proliferating cell nuclear antigen, superoxide dismutase, and ubiquitin, among others.

One of the unique characteristics of baculoviruses is their ability to produce two morphologically and biochemically distinct forms of virus particles. These two different virus morphotypes each play a specific role in the life cycle of the virus. The first to be produced during infection is called *budded virus*, and consists of a single nucleocapsid surrounded by an envelope acquired from the cell plasma membrane during budding. The budded form is infectious in cell culture, and is responsible for transmission of the virus within an infected insect larva. Later in infection the second form, *occluded virus*, is produced within the nucleus of the infected cell. Occluded virus consists of one or many (depending on the particular baculovirus) enveloped nucleocapsids encased within a proteinaceous crystalline matrix primarily consisting of a single polypeptide called *polyhedrin*. The entire protein matrix is called an *occlusion body* and is large enough to be easily seen with the light microscope. The nucleocapsids within the occlusion body acquire their envelopes through an unknown mechanism within the nucleus. The occlusion bodies are stable in the environment and are the form of virus that is responsible for horizontal transmission of the virus from one insect to another through feeding.

In addition to being morphologically distinct, the two types of virus can be biochemically distinguished by their protein composition. Some proteins, such as the capsid protein, are shared by both types, whereas others, such as the envelope proteins, are unique to one form or the other.

2.2. Replication Cycle

The replication cycle of AcMNPV (Fig. 1) begins with the entry of budded virus into a susceptible host cell by endocytosis. The viral genome makes its way to the nucleus, where a set of early genes are transcribed by the host RNA polymerase II. These early mRNAs are translated in the cytoplasm and encode proteins that are involved in the transcription of other early genes and late genes, and in viral DNA replication. In addition, there are a number of auxiliary proteins that are not strictly required for viral gene expression or replication but that provide advantages to the virus, including proteins that are required to prevent cellular apoptosis.

At some point (around 6 hr postinfection for AcMNPV in cultured cells), there is a transition from early gene expression to late gene expression. This transition involves the initiation of viral DNA synthesis via a mechanism that is not well understood. What is known is that viral DNA replication must begin before late genes can be expressed. *In vitro* studies have shown that a total of 6 viral proteins are required for viral DNA synthesis, and an additional 12 proteins are required for transcription from a late promoter. The late genes are transcribed by a different RNA polymerase activity than are the early genes. The late gene-specific RNA polymerase has not been identified to date; it may be virally encoded, or it may be a virally modified host polymerase. The expression of the late genes is important for virus assembly, as late proteins include those that make up the structural components of the virus particles. Later (around 18 hr postinfection for AcMNPV), another shift occurs in transcription to the very late genes, which are

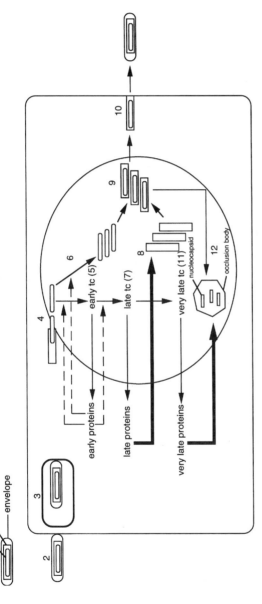

FIGURE 1. A simplified version of the replication cycle of a typical baculovirus. The budded virus particle (1), consisting of the viral DNA contained within the capsid and surrounded by an envelope, attaches to the host cell (2) and is internalized by endocytosis. Once internalized, the viral envelope fuses with the endosome (3), releasing the nucleocapsid into the cytoplasm. The nucleocapsid makes its way to the nucleus, into which the viral DNA is inserted (4). Transcription (tc) of early viral genes occurs (5), and early mRNA transcripts are translated in the cytoplasm into early proteins, some of which have important roles in early transcription, late transcription, and viral DNA synthesis (6). Once DNA replication has begun, late transcription proceeds (7), leading to the synthesis of late proteins, some of which are involved in the construction of new progeny virus (8). Assembly of the nucleocapsids occurs (9) in the nucleus and the nucleocapsids are transported to the plasma membrane, through which they bud (10), acquiring a new envelope in the process which contains viral transmembrane proteins. Eventually, very late transcription (11) produces very late mRNAs and proteins, some of which are involved in assembly of occlusion bodies, proteinaceous crystalline bodies containing nucleocapsids (12). Dashed arrows indicate positive roles in transcription or DNA replication, and thick arrows indicate roles for late and very late structural proteins.

expressed at extremely high levels and encode proteins that are required for occluded virus structure, such as polyhedrin.

The baculoviruses are primarily lytic viruses, meaning that they cause death of the host cell. Normally, AcMNPV infection of the cell line SF-21, derived from the fall armyworm *Spodoptera frugiperda*, results in cell death 3 to 4 days after infection. The type of cell death that is observed at this time appears to be necrotic or passive lysis, in which the cells lose membrane integrity, swell, and burst.

2.3. Induction of Apoptosis by Baculoviruses

The insight that AcMNPV was capable of inducing apoptotic cell death occurred during the study of a spontaneous virus mutant called the annihilator mutant, or vAcAnh (Clem *et al.*, 1991). This mutant caused more rapid death of SF-21 cells than did the wild-type virus, with most of the cells dying within the first 24 hr after infection. In addition to the death being much more rapid, the dramatic blebbing morphology of the dying cells suggested that they were undergoing apoptosis. The mutation in vAcAnh was mapped by marker rescue to a single open reading frame in the viral genome called *p35*. Furthermore, specific mutations introduced into the *p35* gene also resulted in an annihilator phenotype. These results led to the conclusion that the *p35* gene is normally responsible for preventing cell death by apoptosis. The failure to block the apoptotic response has drastic negative effects on viral replication, including a total lack of production of occluded virus, a 100-fold or more reduction in the yield of budded virus, and delays and reductions in the expression of viral genes (Clem and Miller, 1993; Hershberger *et al.*, 1992).

Virus mutants lacking *p35* are also dramatically less infectious in S. *frugiperda* larvae than is wild-type AcMNPV, and production of progeny occluded virus is also severely affected (Clem *et al.*, 1994; Clem and Miller, 1993). Thus, it appears that the apoptotic death of SF-21 cells is an antiviral response that limits the replication and subsequent spread of viruses in the insect host. This hypothesis is further supported by the response of a different insect species to infection by vAcAnh. Larvae and cell lines of the cabbage looper *Trichoplusia ni* are highly susceptible to AcMNPV infection. When vAcAnh or other *p35* mutant viruses are used to infect larvae and cell lines of *T. ni* such as the TN-368 cell line, the mutant viruses have a wild type phenotype (Clem *et al.*, 1994; Clem and Miller, 1993; Hershberger *et al.*, 1992). In other words, no apoptosis is seen in TN-368 cells, and normal levels of budded and occluded virus are produced, and in *T. ni* larvae the mutant viruses are equally as infectious as wild-type AcMNPV. The correlation between a lack of apoptosis in the TN-368 cell line and normal infectivity in *T. ni* larvae further supports the hypothesis that the ability of the infected *S. frugiperda* cells to commit suicide on infection confers a protective advantage to the insect. These results also suggest that acquisition of the *p35* gene has allowed the virus to expand its host range to include *S. frugiperda* (Clem and Miller, 1993). It remains to be formally proven, however, that apoptosis is directly responsible for the lower infectivity of *p35* mutant viruses in *S. frugiperda* larvae.

When SF-21 cells are infected with wild-type AcMNPV, there is a transient blebbing at around 12 hr postinfection that resembles the early stages of apoptosis (Clem *et al.*, 1991), suggesting that infection with the wild-type virus also induces the early stages of apoptosis but that expression of *p35* is able to halt the death process at some point. This transient blebbing is not seen in TN-368 cells infected with either vAcAnh or wild-type AcMNPV. Wild-type AcMNPV also induces apoptosis in the SL2 cell line derived from *Spodoptera littoralis* (Chejanovsky and Gershburg, 1995), a species that is nonpermissive for AcMNPV infection. This may be another interesting case where apoptosis limits the host range of AcMNPV, this time of the wild-type virus.

2.3.1. Apoptosis of SF-21 Insect Cells versus Mammalian Cells

Insect cells have been known to undergo programmed cell death for some time; some of the original descriptions of developmentally programmed cell death are from insect tissues (Lockshin, 1985). However, insect cells had not been conclusively shown to undergo that form of programmed cell death known as apoptosis, although there had been some descriptions of insect cell death with similar morphology (Pipan and Rakovec, 1980; Giorgi and Deri, 1976). The morphology of vAcAnh-infected SF-21 cells led to the investigation of whether or not apoptosis was occurring in these cells, as there was drastic surface blebbing and disintegration of the cells into apoptotic bodies. When the DNA from these cells was examined, it was digested into oligonucleosomal length fragments, and also was condensed within the nucleus (Clem *et al.*, 1991). The mitochondria of the SF-21 cells remained intact throughout the blebbing process, further suggesting that this was a form of active cell death. The blebs that were shed from the dying cells contained both cytoplasmic organelles and condensed chromatin (Clem *et al.*, 1991). Thus, the process of cell death we observe in SF-21 cells appears quite similar to that of mammalian cell apoptosis, with the major difference being the length of time required for an individual cell to undergo the blebbing process (1 to 2 hr for SF-21 cells versus several minutes for most mammalian cells). This may be related at least in part to the lower temperature (27°C) at which SF-21 cells are cultured.

2.3.2. Possible Mechanisms of Baculovirus-Mediated Induction of Apoptosis

The induction of apoptosis in SF-21 cells by vAcAnh infection appears to be associated with the transition from the early to late stages of infection. During this crucial time, several important processes are occurring simultaneously in the infected cell. (1) Viral DNA synthesis is initiated. The transcription and translation of early genes has resulted in the accumulation of proteins that are required for viral DNA replication, and although the trigger for the initiation of DNA synthesis is not known, it may simply be the accumulation of sufficient levels of these proteins and their engagement of the viral DNA. (2) Late gene expression begins. Both inhibitor studies and *in vitro* transfection assays have determined that viral DNA replication is required for the expression of the late genes. By an unknown mechanism, this allows the late RNA polymerase activity to initiate transcription, and simultaneously the transcription activity of the cellular RNA polymerase II enzyme is attenuated. (3) The synthesis of cellular RNA and protein, along with the expression of early viral genes, is gradually shut down by an unknown process(es). By 12 to 18 hr postinfection, there is almost no detectable cellular RNA or protein synthesis, and only late and very late viral products are detected.

When vAcAnh-infected cells are treated with inhibitors that block either both the early and late phases or only the late phase, apoptosis is not triggered (Clem and Miller, 1994a), suggesting that the initiation of apoptosis is somehow related to the transition from the early to late stages of infection. However, it is difficult to determine which of the activities listed above are actually responsible for the induction of apoptosis, because they are so closely linked. It is possible that the cessation of host RNA synthesis is the trigger, as inhibitors of RNA synthesis are potent inducers of apoptosis in these cells (see below). It is also possible that an unplanned round of DNA synthesis triggers the cells to die. When the viral genes necessary for DNA synthesis are expressed in SF-21 cells, very little DNA can be recovered unless *p35* is included, suggesting that apoptosis is triggered by the expression of the viral DNA synthesis machinery (Lu and Miller, 1995). At this point it appears that either stimulus is sufficient to induce apoptosis in SF-21 cells. Recent results indicate that expression of the viral transactivator IE-1 (one of the genes required for viral DNA replication) is sufficient to induce apoptosis on its own (Prikhod'ko and Miller, 1996), so clearly this area of research is still evolving.

2.3.3. Nonviral Inducers of Apoptosis in SF-21 Cells

Three different inhibitors of RNA synthesis, each of which work by a different mechanism (actinomycin D, which intercalates into DNA and prevents elongation; α-amanitin, which binds to and inactivates the large subunit of RNA polymerase II; and the nucleoside analogue 5, 6-dichlorobenzimidazole riboside), have been shown to induce rapid and widespread apoptosis in SF-21 cells (Clem and Miller, 1994a). Although RNA and protein synthesis are often required for the active process of apoptosis, inhibition of these processes has been shown to induce apoptotic death in some cell types (Raff *et al.*, 1993). Presumably, the apoptotic machinery is already in place in these cells and the continued synthesis of an inhibitory protein is required to prevent the initiation of death. Interestingly, cycloheximide does not induce apoptotic death in SF-21 cells, despite the fact that it is an effective protein synthesis inhibitor in this cell line (Clem and Miller, 1994a). It is possible that the synthesis of a specific RNA species is required to inhibit death, although it is more difficult to imagine how an RNA molecule would function to block apoptosis. Perhaps it is the process of transcription itself that is required, but it seems more likely that a protein is responsible. It may simply be that cycloheximide is a more leaky drug than the RNA synthesis inhibitors tested, and only low levels of the inhibitory protein are required for function.

In the SF-9 cell line (a clonal derivative of SF-21), the phosphatase inhibitor okadaic acid has also been shown to induce apoptosis (Bergqvist and Magnusson, 1994), although in our hands it is less effective than actinomycin D (unpublished results). Okadaic acid is purported to be an inhibitor of protein phosphatase 2A, but it also has effects on other phosphatases. Thus, protein phosphatases may play a role in promoting the survival of SF-21 cells, but at this time it is not possible to say with confidence which phosphatases are important.

Although gamma and ultraviolet radiation are potent inducers of apoptotic death in many mammalian cell types, insect cells, particularly those of lepidopterans, are much more resistant to radiation than are mammalian cells (Koval, 1996). High doses of radiation will kill lepidopteran cells, but whether the death is apoptotic has not been carefully examined. In mammals, the tumor suppressor gene *p53* is normally required for the induction of apoptosis by DNA-damaging agents such as radiation (Strasser *et al.*, 1994; Clarke *et al.*, 1993; Lowe *et al.*, 1993); it must be noted that no *p53* homologue has been discovered in any invertebrate to date, despite exhaustive searches. The ability of invertebrates to tolerate higher levels of radiation may be related to more efficient DNA repair mechanisms or other strategies (Koval, 1996). Interestingly, TN-368 cells are even more resistant to the damaging effects of radiation than other lepidopteran cell lines. In fact, the TN-368 cell line appears to be highly resistant to the induction of apoptosis by a variety of stimuli, including vAcAnh infection, RNA synthesis inhibitor treatment, and irradiation. However, apoptotic death of TN-368 cells can be induced by expression of the mammalian ICE gene (Bump *et al.*, 1995). Thus, the final effector pathway appears to be intact in TN-368 cells, but there may be either a defect in an apoptotic signaling pathway or an overexpressed protective gene that makes these cells extraordinarily resistant to stimuli that normally result in the death of other cells.

2.4. Baculovirus Genes that Protect Cells against Apoptosis

2.4.1. The p35 Gene

2.4.1a. Characteristics of the p35 Protein.   The *p35* gene has no detectable homology to other known genes, except for the *p35* homologue present in the closely related baculovirus *Bombyx mori*

NPV, which is 90% identical to AcMNPV *p35* at the amino acid level and can also prevent apoptosis induced by BmNPV in *B. mori* cells (Kamita *et al.*, 1993). The protein encoded by AcMNPV *p35* is 299 amino acids in length and has a predicted molecular mass of 34.8 kDa. The protein has no known sequence motifs, and lacks a signal sequence. It appears to be a cytoplasmic protein (Hershberger *et al.*, 1994). The most remarkable features of the *p35* protein are lysine-rich domains at the center of the molecule and at its carboxyl terminus, and a highly charged domain between residues 64 and 100 (20 out of 37 amino acids being acidic or basic) (Clem *et al.*, 1996). These charged domains would be expected to be exposed on the outer surface of the molecule.

The carboxyl terminus of P35 may be important for its function, as mutations in this region abolish antiapoptotic activity (Hershberger *et al.*, 1992). However, the effect of these mutations on protein stability has not been examined. The amino-terminal portion of the molecule can serve as a dominant negative inhibitor of P35 function (Cartier *et al.*, 1994). When a construct expressing amino acids 1 to 76 was stably expressed in SF-21 cells, the cells still underwent apoptosis on wild-type AcMNPV infection, despite the expression of full-length P35 protein. There was little or no accumulation of full-length *p35* protein in these infected cells, despite the fact that other viral proteins accumulated to normal levels, suggesting that the interaction of full-length P35 with its amino-terminal domain may target the full-length polypeptide for rapid degradation. This, along with yeast two-hybrid results (Rohrmann and Leisy, 1995), indicates that *p35* may form homodimers. The entire *p35* protein is extremely sensitive to any insertions or deletions, as almost any insertion or deletion in the molecule abolishes or greatly reduces antiapoptotic function.

2.4.1b. P35 Is an Inhibitor of ICE-like Proteases. It has recently been shown that the P35 polypeptide potently inhibits several members of the interleukin-1 beta converting enzyme (ICE) family of cysteine proteases (Bump *et al.*, 1995; Xue and Horvitz, 1995). A large body of data now implicates the ICE proteases in apoptotic cell death (reviewed in Martin and Green, 1995). The inhibition of ICE-like proteases by P35 has been demonstrated both *in vitro* and in cells. When expressed in COS cells, P35 inhibits the cleavage of pro-IL-1B by ICE (Bump *et al.*, 1995). *In vitro*, purified P35 can efficiently block the activity of purified ICE, ICH-1, ICH-2, and CPP32 (Bump *et al.*, 1995), as well as the nematode ICE homologue CED-3 (Xue and Horvitz, 1995). The ability of P35 to inhibit the mammalian enzymes at a 1:1 molar ratio (Bump *et al.*, 1995) indicates that it is acting as an irreversible inhibitor rather than a competitive substrate.

P35 is cleaved by ICE into two fragments, which then form a stable complex with the enzyme (Bump *et al.*, 1995). The cleavage occurs between residues 87 and 88 of P35, after an aspartic acid residue, which is the characteristic cleavage site for the ICE-like proteases. This cleavage is essential for both inhibition and complex formation (Bump *et al.*, 1995). The cleavage site is within the highly charged region of the protein, and thus is predicted to be exposed on the outer surface.

P35 differs from another viral inhibitor of ICE, the cowpox crmA protein, in several important ways. Whereas crmA is a member of the serpin family of protease inhibitors, P35 has no homology to serpins. P35 is also unable to inhibit granzyme B, a serine protease that also cleaves after aspartate residues and is involved in the induction of apoptotic death by cytotoxic T cells, whereas crmA does inhibit this enzyme (Quan *et al.*, 1995). Thus, although P35 and crmA act in an analogous fashion, there are likely to be important differences in their interactions with ICE and further study of these interactions should yield interesting and novel information.

2.4.2. The iap Gene Family

2.4.2a. The Discovery of Baculovirus iap. Once it was clear that AcMNPV contained a gene that could block apoptosis, the vAcAnh mutant was used to search for antiapoptotic genes in other baculoviruses (Birnbaum *et al.*, 1994; Crook *et al.*, 1993). A genetic screen was used that took advantage of the fact that vAcAnh-infected SF-21 cells do not produce occlusion bodies. By cotransfecting vAcAnh DNA and genomic DNA from other baculoviruses into SF-21 cells, antiapoptotic genes were identified by looking for the presence of occlusion bodies in the transfected cells several days after transfection. A limited screen of several other baculoviruses resulted in the discovery of a second family of antiapoptotic genes, called *iap* (*i*nhibitor of *ap*optosis). Three different baculoviruses are known to contain *iap* genes. The genomic DNA from the baculoviruses *Cydia pomonella* granulosis virus (CpGV) and *Orgyia pseudotsugata* NPV (OpMNPV) was able to rescue occlusion body formation when cotransfected into SF-21 cells with vAcAnh DNA, and further analysis led to the identification of Cp-*iap* and Op-*iap*, respectively (Birnbaum *et al.*, 1994; Crook *et al.*, 1993). A third baculovirus *iap* gene was identified in the AcMNPV genome by sequence homology, but the AcMNPV *iap* gene (Ac-*iap*) does not appear to have antiapoptotic activity (Clem and Miller, 1994a).

    2.4.2b. Sequence Motifs Found in IAP Proteins. The *iap* genes have no homology to *p35* and are characterized by distinct sequence motifs (Fig. 2) including a type of zinc finger known as a RING finger, which is found at their carboxyl termini, and two novel imperfect repeats near their amino termini known as baculovirus IAP repeats (BIRs) (Birnbaum *et al.*, 1994). Both types of motifs are predicted to be metal coordinating sequences.

    The RING finger is a type of zinc finger motif that was first described in 1991 (Freemont *et al.*, 1991). Around 50 proteins with a wide variety of functions are now known to contain a RING finger. To date, sequence-specific DNA binding has not been demonstrated for any RING finger motif, and they may be important in protein–protein interactions (Berg and Shi, 1996). Solution structures of two RING fingers have been examined, and in both cases the motif was found to bind two atoms of zinc (Borden *et al.*, 1995; Barlow *et al.*, 1994). Although the basic core structure of the two RING fingers was the same, there were significant differences in their overall three-dimensional structures, which could help explain their diversity of functions.

    BIRs are repeated sequences found near the amino termini of IAP proteins (Fig. 2). An individual BIR is approximately 65 amino acids in length and has a conserved $CysX_2Cys$ motif near its center followed by a $HisX_6Cys$ motif near its carboxyl end (where X is any amino acid) (Clem *et al.*, 1996). In addition, there are a number of other highly conserved residues, including six residues that are absolutely conserved in all known BIRs (Fig. 2). The BIRs are only found in a very limited set of proteins, including the IAPs, NAIP (see below), an incomplete open reading frame from Chilo iridescent virus (Crook *et al.*, 1993), and a structural protein called A224L from African swine fever virus (Chacon *et al.*, 1995).

    2.4.2c. Cellular Homologues of Baculovirus iap. Although the mode of action of the IAP proteins is unknown, recent exciting progress has been made in this area through the identification of cellular IAP homologues. The first of these identified genes was

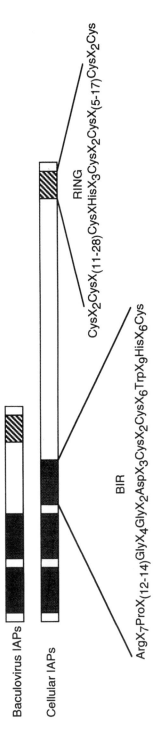

FIGURE 2. A schematic representation of the primary structure of the IAP proteins, both baculoviral and cellular. The shaded boxes represent the BIRs, and the hatched boxes the RING fingers. Below are shown the consensus sequences of the BIR and RING finger motifs, with X representing any amino acid. Note that only absolutely conserved residues are shown here; for more comprehensive consensus sequences see Clem *et al.* (1996) and Clem and Miller (1994b).

Baculovirus IAPs

Cellular IAPs

BIR

$ArgX_7ProX_{(12-14)}GlyX_4GlyX_2AspX_3CysX_2CysX_6TrpX_9HisX_6Cys$

RING

$CysX_2CysX_{(11-28)}CysXHisX_3CysX_2CysX_{(5-17)}CysX_2Cys$

NAIP, which encodes a human protein frequently mutated in patients with spinal muscular atrophy, a neurodegenerative disease that affects infants (Roy *et al.*, 1995). The NAIP protein has three BIRs at its amino terminus but no other detectable homology to the IAPs. At 140 kDa, it is a much larger molecule than the baculovirus IAPs, which are around 30 to 35 kDa in mass. Three other human proteins have been identified that appear to be closer relatives to the baculovirus IAPs (Duckett *et al.*, 1996; Liston *et al.*, 1996; Uren *et al.*, 1996; Rothe *et al.*, 1995). These proteins, cIAP-1/hIAP-2/MIHB, cIAP-2/hIAP-1/MIHC, and hILP/X-IAP/MIHA, also have three rather than two BIRs, but in addition contain a RING finger at their carboxyl termini (Fig. 2). They also have approximately 150 amino acids between the third BIR and the RING finger not found in the baculovirus IAPs.

The proteins encoded by cIAP-1 and cIAP-2 interact with TNF receptor-associated factors (TRAFs) 1 and 2 both *in vitro* and in cells (Uren *et al.*, 1996; Rothe *et al.*, 1995). The BIRs are necessary and sufficient for this interaction (Rothe *et al.*, 1995). However, the importance of this interaction in protecting against apoptosis is unclear. There have been conflicting reports on whether or not cIAP-1 and 2 can protect cells against apoptotic death (Liston *et al.*, 1996; Uren *et al.*, 1996; R. Clem and J. M. Hardwick, unpublished results). hILP, on the other hand, does seem to have strong antiapoptotic activity (Duckett *et al.*, 1996; Liston *et al.*, 1996; Uren *et al.*, 1996) but does not interact with TRAFs 1 and 2 (Uren *et al.*, 1996; C. Duckett and C. Thompson, personal communication).

The *Drosophila* genome also contains IAP-homologous proteins (Duckett *et al.*, 1996; Uren *et al.*, 1996; Hay *et al.*, 1995). Two proteins have been described, DIAP1 and DIAP2 (Hay *et al.*, 1995). The DIAP1 protein is the product of the *thread* locus. There are several allelic mutants of *thread*, most of which are lethal (Hay *et al.*, 1995). Overexpression of either DIAP1 or 2 is able to partially prevent cell death in the developing eye, both normally occurring cell death and death induced by expression of *rpr* or *hid*. Both DIAP1 and 2 are similar in structure to the other known IAPs, with DIAP1 containing two BIRs and DIAP2 containing three BIRs, and both proteins having a RING finger at their carboxyl termini (Hay *et al.*, 1995).

2.4.2d. Functional Domains of IAP. Although attempts have been made to define the domains of IAPs that are important for blocking cell death, the roles of these domains have yet to be precisely defined. When the BIR or RING finger domain of Cp-IAP and the nonfunctional homologue Ac-IAP were exchanged, it was found that hybrid proteins containing either the BIRs or RING finger of Ac-IAP were inactive, suggesting that both domains are important for function (Clem and Miller, 1994a). Cp-IAP constructs lacking only the RING finger were also inactive (R. Clem and L. Miller, unpublished results). However, when a version of DIAP1 lacking the RING finger was expressed in the developing eye, it still prevented cell death equally as well as the full-length protein, and actually blocked *hid*-dependent death more efficiently than the full-length protein (Hay *et al.*, 1995). Similarly, a version of human cIAP-1 lacking the RING finger was able to partially block normally occurring cell death in the *Drosophila* eye, whereas the full-length cIAP-1 protein was inactive (Hay *et al.*, 1995). Expression of a DIAP1 gene lacking the BIRs appeared to induce extra cell death in the developing eye (Hay *et al.*, 1995). Taken together, these results suggest that the BIRs are crucial for antiapoptotic function of the IAPs, and that the role of the RING finger, at least in the cellular IAP homologues, may be to negatively regulate the activity of the protein. However, the results obtained with Cp-IAP do not support the latter conclusion. Clearly, further studies are needed to delineate the specific roles of these important motifs.

3. WHAT DO BACULOVIRUSES TELL US ABOUT CELL DEATH PATHWAYS IN OTHER ORGANISMS?

The study of baculovirus regulation of cell death has led to some of the most convincing evidence to date that cell death pathways are conserved in metazoans ranging from insects and nematodes to humans (Fig. 3). The polypeptide encoded by the *p35* gene is the only antiapoptotic protein that has been convincingly shown to block cell death in insects, nematodes, and mammals. The ability of baculovirus *iap* to block apoptosis in mammals and the presence of *iap* homologues in both insects and mammals is further evidence that the basic cell death pathways are similar in these evolutionarily distant organisms.

3.1. Nematodes

Study of the nematode *Caenorhabditis elegans* has provided one of the most complete genetic cell death pathways in any organism to date (reviewed in Ellis *et al.*, 1991). The precise position and

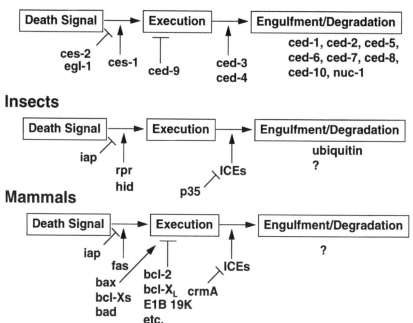

FIGURE 3. A comparison of the genetic pathways controlling apoptosis in nematodes, insects, and mammals. Arrows denote positive regulation, negative regulation is indicated by "T"-shaped symbols. This figure is not intended to be a comprehensive listing of all of the known genes involved in cell death regulation, of which there are now many more (particularly mammalian), but is only designed to highlight the similarities between the pathways in these different organisms.

fate of all of the cells in this animal are known, making it possible to isolate mutants that affect cell death during development. During normal larval development, 131 cells die by programmed cell death. At least two genes are required for this process, *ced*-3 and *ced*-4. Whereas *ced*-4 encodes a protein with no known mammalian homologues (Yuan and Horvitz, 1992), *ced*-3 encodes an ICE family member (Yuan *et al.*, 1993). Overexpression of *ced*-3 in transgenic nematodes results in the inappropriate death of a large number of cells that would normally live. Countering the action of *ced*-3 and *ced*-4 is the *ced*-9 gene, which is required for cells to survive (Hengartner *et al.*, 1992), and which is a homologue of the mammalian *bcl*-2 gene (Hengartner and Horvitz, 1994). Overexpression of *ced*-9 prevents the death of all of the cells that would normally die during development. Mutants lacking *ced*-9 resemble the phenotype of animals overexpressing *ced*-3 or *ced*-4, in which massive inappropriate cell death occurs. In addition to these three important genes, a number of other genes which have defined roles in the degradation and phagocytosis of the dead and dying cells (Ellis *et al.*, 1991).

The ability to construct transgenic nematodes that express foreign genes has been a useful technique for providing information concerning the conservation of death pathways in this simple organism and higher animals. Overexpression of human *bcl*-2 (Hengartner and Horvitz, 1994; Vaux *et al.*, 1992) or baculovirus *p35* (Xue and Horvitz, 1995; Sugimoto *et al.*, 1994) in transgenic nematodes can prevent some of the deaths that occur during development, thus demonstrating conserved roles for these anti-apoptotic genes. In hindsight, these results are not surprising given that *bcl*-2 has homology to *ced*-9, and *p35* is an inhibitor of ICE proteases, including *ced*-3 (Xue and Horvitz, 1995). In an interesting example of the elegance of the gene transfer approach (Xue and Horvitz, 1995), the ICE inhibitor crmA, which was unable to prevent cell death in *C. elegans*, was converted to a functional protein by the addition of the ICE cleavage site found in P35. This suggests that P35 may have a broader ability to interact with ICE proteases than does crmA; however, unlike crmA, P35 does not inhibit granzyme B. Because CPP32/apopain (which is postulated to be the death protease) and CED-3 are in the same subfamily of ICE-like enzymes (Nicholson *et al.*, 1995), it may be that P35 is able to interact with this subfamily most effectively, whereas crmA may have evolved to inhibit the subfamily containing ICE itself.

3.2. Insects

Although less defined than the death pathways of *C. elegans*, our understanding of the genetic pathways of insect cell death has made considerable progress in recent years. In *Drosophila*, proper development of the eye requires the death of a number of cells in the retina (Wolff and Ready, 1991). The *Drosophila* gene *reaper* (*rpr*) encodes a 65-amino-acid polypeptide that is involved in programmed cell death in the eye as well as in the rest of the developing embryo (White *et al.*, 1994). In mutants lacking the chromosomal region containing *rpr*, there is suppression of all of the normally occurring programmed cell death during development of the embryo, whereas transgenic flies that can be induced to overexpress *rpr* from a heat shock promoter exhibit massive cell death throughout the embryo (and the flies die as a result). Overexpression of *rpr* using a promoter specifically expressed in the developing eye results in extensive inappropriate cell death and eye ablation (White *et al.*, 1996). Similarly, *rpr* overexpression in Schneider's *Drosophila* (S2) cell line results in apoptotic death, which can be blocked by a peptide inhibitor of ICE-like proteases (Pronk *et al.*, 1996). The *rpr* gene has limited homology to the cytoplasmic death domain of the mammalian tumor necrosis factor receptor family (Golstein *et al.*, 1995), including *fas*, a molecule that signals apoptosis (Nagata and Golstein, 1995). However, whereas *fas* is anchored in the plasma membrane, *rpr* lacks the membrane-spanning and extracellular domains of *fas*. Thus, *rpr* may represent an ancestral form of the *fas* antigen that is strictly intracellular but that plays a similar role in a death signaling pathway.

A second *Drosophila* gene, *head involution defective* (*hid*), seems to play a role similar to that of *rpr* (Grether *et al.*, 1995). The two genes appear to be functionally redundant, as deletion of both genes (they lie adjacent to each other in the fly genome) is required for the *rpr* mutant phenotype and overexpression of either gene can cause inappropriate death in the developing eye or other tissues of the embryo (White *et al.*, 1996; Grether *et al.*, 1995). The *hid* gene encodes a 410-amino-acid protein with no homology to other known proteins.

The *p35* gene is able to block cell death in *Drosophila*, including normally occurring death, death induced by x-irradiation (Hay *et al.*, 1994) or death induced by overexpression of *hid* (Grether *et al.*, 1995) or *rpr* (White *et al.*, 1996). Baculovirus *iap* is also able to block *rpr*-induced death in the *Drosophila* eye, although not nearly as well as *p35* (Hay *et al.*, 1995). The *Drosophila iap* homologues DIAP-1 and DIAP-2 are comparable to *p35* in their ability to block both normally occurring death and death induced by *rpr* or *hid* (Hay *et al.*, 1995).

The ability of mammalian genes to function in insects is varied. For example, the *bcl-2* gene does not appear to have any antiapoptotic function in insects. Expression of *bcl-2* in the developing *Drosophila* eye does not block cell death (B. Hay and G. Rubin, personal communication), nor does *bcl-2* expression prevent apoptosis in SF-21 cells infected with *p35* mutants of Ac-MNPV or treated with actD (Cartier *et al.*, 1994; Clem and Miller,

1994a). A report of antiapoptotic function for *bcl*-2 at very late times in baculovirus infection (Alnemri *et al.*, 1992) may have been the result of the ability of *bcl*-2 to also prevent necrotic cell death (Kane *et al.*, 1993). On the other hand, expression of human ICE-like proteases in SF-21 or SF-9 cells results in apoptosis (Bump *et al.*, 1995; Fernandes-Alnemri *et al.*, 1994, 1995), and poxvirus *crmA* is able to block apoptosis stimulated by some (but not all) ICE-like proteases in SF-21 cells (S. Seshagiri and L. Miller, personal communication). The *crmA* gene is also able to partially block *rpr*-induced death in SF-21 cells, although not as well as *p35* (D. Vucic, S. Seshagiri, and L. Miller, personal communication).

3.3. Mammals

The ability of *p35* to block apoptosis in mammalian cells is well documented. Expression of *p35* is able to inhibit apoptosis of mammalian neural cells stimulated by glucose withdrawal, serum withdrawal, or calcium ionophore (Rabizadeh *et al.*, 1993). Expression of *p35* in sympathetic neurons also protects them from nerve growth factor deprivation (Martinou *et al.*, 1995). Breast carcinoma cells are also protected by *p35* expression from death induced by tumor necrosis factor or *fas* (Beidler *et al.*, 1995). Finally, the *p35* gene also protects hamster fibroblasts and murine neuroblastoma cells from Sindbis virus-induced apoptosis (V. Nava, R. Clem, and J. M. Hardwick, unpublished results). The ability of this insect virus gene to function in such a wide variety of situations strongly suggests that there is a highly conserved role for ICE-like proteases in cell death.

Preliminary results suggest that the ability of baculovirus *iap* to block apoptosis in mammalian cells appears to be more restricted than that of *p35* (Clem *et al.*, 1996). This suggests that *iap* may act at a position farther upstream in the cell death pathway than *p35*, prior to the point where multiple signaling pathways converge. The mechanism of *iap* function remains to be elucidated, but the discovery of mammalian *iap* homologues promises to speed the pace of discovery.

4. CONCLUDING REMARKS

Study of the induction and inhibition of apoptosis by baculoviruses has greatly increased our understanding of the genetic pathways of cell death not only in insects but in higher animals as well. For example, if not for the discovery of the antiapoptotic function of the baculovirus *iap* genes, our present understanding of the potential role of NAIP in protecting neurons from spinal muscular atrophy would quite possibly still be years away. In addition, the ability of *p35* to block cell death in nematodes, insects, and mammals is the best indication so far that there is a conserved role for ICE-like proteases in the death programs of these diverse organisms. Finally, analysis of the replication and infectivity of mutant viruses lacking *p35* has also provided the strongest evidence to date that apoptosis can act as an antiviral response in lower animals lacking a sophisticated immune system. The ability of individual cells to commit suicide in response to stress signals is undoubtedly vital to the survival of the organism, and may in part explain why apoptosis is so widely conserved in evolution.

REFERENCES

Alnemri, E. S., Robertson, N. M., Fernandes, T. F., Croce, C. M., and Litwack, G., 1992, Overexpressed full-length human BCL2 extends the survival of baculovirus-infected Sf9 insect cells, *Proc. Natl. Acad. Sci. USA* **89:**7295–7299.

Ayres, M. D., Howard, S. C., Kuzio, J., Lopez-Ferber, M., and Possee, R. D., 1994, The complete DNA sequence of *Autographa californica* nuclear polyhedrosis virus, *Virology* **202:**586–605.

Barlow, P. N., Luisi, B., Milner, A., Elliott, M., and Everett, R., 1994, Structure of the C_3HC_4 domain by ^{1}H-nuclear magnetic resonance spectroscopy, *J. Mol. Biol.* **2337:**201–211.

Beidler, D. R., Tewari, M., Friesen, P. D., Poirier, G., and Dixit, V. M., 1995, The baculovirus *p35* protein inhibits fas- and tumor necrosis factor-induced apoptosis, *J. Biol. Chem.* **270:**16256–16258.

Berg, J. M., and Shi, Y., 1996, The galvanization of biology: A growing appreciation for the roles of zinc, *Science* **271:**1081–1085.

Bergqvist, A., and Magnusson, G., 1994, Apoptosis of *Spodoptera frugiperda* cells induced by okadaic acid is abrogated by baculovirus infection, *Exp. Cell Res.* **215:**223–227.

Birnbaum, M. J., Clem, R. J., and Miller, L. K., 1994, An apoptosis-inhibiting gene from a nuclear polyhedrosis virus encoding a peptide with cys/his sequence motifs, *J. Virol.* **68:**2521–2528.

Borden, K. L. B., Boddy, M. N., Lally, J., O'Reilly, N. J., Martin, S., Howe, K., Solomon, E., and Freemont, P. S., 1995, The solution structure of the RING finger domain from the acute promyelocytic leukaemia proto-oncoprotein PML, *EMBO J.* **14:**1532–1541.

Bump, N. J., Hackett, M., Hugunin, M., Seshagiri, S., Brady, K., Chen, P., Ferenz, C., Franklin, S., Ghayur, T., Li, P., Licari, P., Mankovich, J., Shi, L., Greenberg,

A. H., Miller, L. K., and Wong, W. W., 1995, Inhibition of ICE family proteases by baculovirus antiapoptotic protein *p35*, *Science* **269**:1885–1888.

Cartier, J. L., Hershberger, P. A., and Friesen, P. D., 1994, Suppression of apoptosis in insect cells stably transfected with baculovirus *p35*: Dominant interference by N-terminal sequences p35[1-76], *J. Virol.* **68**:7728–7737.

Chacon, M. R., Almazan, F., Nogal, M. L., Vinuela, E., and Rodriguez, J. F.,1995, The African swine fever virus IAP homolog is a late structural polypeptide, *Virology* **214**:670–674.

Chejanovsky, N., and Gershburg, E., 1995, The wild-type *Autographa californica* nuclear polyhedrosis virus induces apoptosis of *Spodoptera littoralis* cells, *Virology* **209**:519–525.

Clarke, A. R., Purdie, C. A., Harrison, D. J., Morris, R. G., Bird, C. C., Hooper, M. L., and Wyllie, A. H., 1993, Thymocyte apoptosis induced by p53-dependent and independent pathways, *Nature* **362**:849–852.

Clem, R. J., and Miller, L. K., 1993, Apoptosis reduces both the in vitro replication and the in vivo infectivity of a baculovirus, *J. Virol.* **67**:3730–3738.

Clem, R. J., and Miller, L. K., 1994a, Control of programmed cell death by the baculovirus genes *p35* and *iap*, *Mol. Cell Biol.* **14**:5212–5222.

Clem, R. J., and Miller, L. K., 1994b, Induction and inhibition of apoptosis by insect viruses, in: *Apoptosis II: The Molecular Basis of Cell Death* (F. O. Cope and L. D. Tomei, eds.), Cold Spring Harbor Laboratory Press, Cold Spring Harbor, NY, pp. 89–110.

Clem, R. J., Fechheimer, M., and Miller, L. K., 1991, Prevention of apoptosis by a baculovirus gene during infection of insect cells, *Science* **254**:1388–1390.

Clem, R. J., Robson, M., and Miller, L. K., 1994, Influence of infection route on the infectivity of baculovirus mutants lacking the apoptosis-inhibiting gene *p35* and the adjacent gene *p94*, *J. Virol.* **68**:6759–6762.

Clem, R. J., Hardwick, J. M., and Miller, L. K., 1996, Anti-apoptotic genes of baculoviruses, *Cell Death Differ.* **3**:9–16.

Clouston, W. M., and Kerr, J. F. R., 1985, Apoptosis, lymphocytotoxicity and the containment of viral infections, *Med. Hypoth.* **18**:399–404.

Crook, N. E., Clem, R. J., and Miller, L. K., 1993, An apoptosis-inhibiting baculovirus gene with a zinc finger-like motif, *J. Virol.* **67**:2168–2174.

Duckett, C. S., Nava, V. E., Gedrich, R. W., Clem, R. J., Van Dongen, J. L., Gilfillan, M. C., Shiels, H., Hardwick, J. M., and Thompson, C. B., 1996, A conserved family of cellular genes related to the baculovirus *iap* gene and encoding apoptosis inhibitors, *EMBO J.* **15**:2685–2694.

Ellis, R. E., Yuan, J., and Horvitz, H. R., 1991, Mechanisms and functions of cell death, *Annu. Rev. Cell Biol.* **7**:663–698.

Fernandes-Alnemri, T., Litwack, G., and Alnemri, E. S., 1994, CPP32, a novel human apoptotic protein with homology to *Caenorhabditis elegans* cell death protein *ced-3* and mammalian interleukin-1beta-converting enzyme, *J. Biol. Chem.* **269**:30761–30764.

Fernandes-Alnemri, T., Litwack, G., and Alnemri, E. S., 1995, *Mch2*, a new member of the apoptotic *Ced-3/Ice* cysteine protease gene family, *Cancer Res.* **55**:2737–2742.

Freemont, P. S., Hanson, I. M., and Trowsdale, J., 1991, A novel cysteine-rich sequence motif, *Cell* **64**:483–484.

Giorgi, F., and Deri, P., 1976, Cell death in ovarian chambers of *Drosophila melanogaster*, *J. Embryol. Exp. Morphol.* **35**:521–533.

Golstein, P., Marguet, D., and Depraetere, V., 1995, Homology between reaper and the cell death domains of Fas and TNFR1, *Cell* **81**:185–186.

Grether, M. E., Abrams, J. M., Agapite, J., White, K., and Steller, H., 1995, The *head involution defective* gene of *Drosophila melanogaster* functions in programmed cell death, *Genes Dev.* **9**:1694–1708.

Hay, B. A., Wolff, T., and Rubin, G. M., 1994, Expression of baculovirus P35 prevents cell death in *Drosophila, Development* **120**:2121–2129.

Hay, B. A., Wassarman, D. A., and Rubin, G. M., 1995, *Drosophila* homologs of baculovirus inhibitor of apoptosis proteins function to block cell death, *Cell* **83**:1253–1262.

Hengartner, M. O., and Horvitz, H. R., 1994, *C. elegans* cell survival gene *ced*-9 encodes a functional homolog of the mammalian proto-oncogene *bcl*-2, *Cell* **76**:665–676.

Hengartner, M. O., Ellis, R. E., and Horvitz, H. R., 1992, *Caenorhabditis elegans* gene *ced*-9 protects cells from programmed cell death, *Nature* **356**:494–499.

Hershberger, P. A., Dickson, J. A., and Friesen, P. D., 1992, Site-specific mutagenesis of the 35-kilodalton protein gene encoded by *Autographa californica* nuclear polyhedrosis virus: Cell line-specific effects on virus replication, *J. Virol.* **66**:5525–5533.

Hershberger, P. A., LaCount, D. J., and Friesen, P. D., 1994, The apoptotic suppressor P35 is required early during baculovirus replication and is targeted to the cytosol of infected cells, *J. Virol.* **68**:3467–3477.

Kamita, S. G., Majima, K., and Maeda, S., 1993, Identification and characterization of the *p35* gene of *Bombyx mori* nuclear polyhedrosis virus that prevents virus-induced apoptosis, *J. Virol.* **67**:455–463.

Kane, D. J., Sarafian, T. A., Anton, R., Hahn, H., Gralla, E. B., Valentine, J. S., Ord, T., and Bredesen, D. E., 1993, Bcl-2 inhibition of neural death: Decreased generation of reactive oxygen species, *Science* **262**:1274–1277.

Kerr, J. F. R., and Harmon, B. V., 1991, Definition and incidence of apoptosis: An historical perspective, in: *Apoptosis: The Molecular Basis of Cell Death* (L. D. Tomei and F. O. Cope, eds.), Cold Spring Harbor Laboratory Press, Cold Spring Harbor, NY, pp. 5–29.

Kerr, J. F. R., Wyllie, A. H., and Currie, A. R., 1972, Apoptosis: A basic biological phenomenon with wide-ranging implications in tissue kinetics, *Br. J. Cancer* **26**:239–257.

Kool, M. and Vlak, J. M., 1993, The structural and functional organization of the *Autographa californica* nuclear polyhedrosis virus genome, *Arch. Virol.* **130**:1–16.

Kool, M., Ahrens, J. M., and Rohrmann, G. F., 1995, Replication of baculovirus DNA, *J. Gen. Virol.* **76**:2103–2118.

Koval, T. M., 1996, Moths: Myths and mysteries of stress resistance, *BioEssays* **18**:149–156.

Liston, P., Roy, N., Tamai, K., Lefebvre, C., Baird, S., Cherton-Horvat, G., Farahani, R., McLean, M., Ikeda, J.-E., MacKenzie, A., and Korneluk, R. G., 1996, Suppression of apoptosis in mammalian cells by NAIP and a related family of IAP genes, *Nature* **379**:349–353.

Lockshin, R. A., 1985, Programmed cell death, in: *Comprehensive Insect Physiology, Biochemistry and Pharmacology* (G. A. Kerkut and L. I. Gilbert, eds.), Pergamon Press, Elmsford, NY, Vol. 2, pp. 301–317.

Lowe, S. W., Schmitt, E. M., Smith, S. W., Osborne, B. A., and Jacks, T., 1993, p53 is required for radiation-induced apoptosis in mouse thymocytes, *Nature* **362**:847–849.

Lu, A., and Miller, L. K., 1995, The roles of eighteen baculovirus late expression factor genes in transcription and DNA replication, *J. Virol.* **69**:975–982.

Martin, S. J., and Green, D. R., 1995, Protease activation during apoptosis: Death by a thousand cuts? *Cell* **82**:349–352.

Martinou, I., Fernandez, P.-A., Missotten, M., White, E., Allet, B., Sadoul, R., and Martinou, J.-C., 1995, Viral proteins E1B19K and *p35* protect sympathetic neurons from cell death induced by NGF deprivation, *J. Cell Biol.* **128**:201–208.

Martz, E., and Howell, D. M., 1989, CTL: Virus control cells first and cytolytic cells second? *Immunol. Today* **10**:79–86.

Miller, L. K., 1995, Genetically engineered insect virus pesticides: Present and future, *J. Invert. Pathol.* **65**:211–216.

Nagata, S., and Golstein, P., 1995, The Fas death factor, *Science* **267**:1449–1456.

Nicholson, D. W., Ali, A., Thornberry, N. A., Vaillancourt, J. P., Ding, C. K., Gallant, M., Gareau, Y., Griffin, P. R., Labelle, M., Lazebnik, Y. A., Munday, N. A., Raju, S. M., Smulson, M. E., Yamin, T. T., Yu, V. L., and Miller, D. K., 1995, Identification and inhibition of the ICE/CED-3 protease necessary for mammalian apoptosis, *Nature* **376**:37–43.

Oltvai, Z. N., and Korsmeyer, S. J., 1994, Checkpoints of dueling dimers foil death wishes, *Cell* **79**:189–192.

O'Reilly, D. R., Miller, L. K., and Luckow, V. A., 1992, *Baculovirus Expression Vectors: A Laboratory Manual*, Freeman, San Francisco.

Pipan, N., and Rakovec, V., 1980, Cell death in the midgut epithelium of the worker honey bee (*Apis mellifera carnica*) during metamorphosis, *Zoomorphologie* **94**:217–224.

Prikhod'ko, E. A., and Miller, L. K., 1996, Induction of apoptosis by baculovirus transactivator IE-1, *J. Virol.* **70**:7116–7124.

Pronk, G. J., Ramer, K., Amiri, P., and Williams, L. T., 1996, Requirement of an ICE-like protease for induction of apoptosis and ceramide generation by REAPER, *Science* **271**:808–810.

Quan, L. T., Caputo, A., Bleackley, R. C., Pickup, D. J., and Salvesen, G. S., 1995, Granzyme B is inhibited by the cowpox serpin cytokine response modifier A, *J. Biol. Chem.* **270**:10377–10379.

Rabizadeh, S., LaCount, D. J, Friesen, P. D., and Bredesen, D. E., 1993, Expression of the baculovirus *p35* gene inhibits mammalian neural cell death, *J. Neurochem.* **61**:2318–2321.

Raff, M. C., Barres, B. A., Burne, J. F., Coles, H. S., Ishizaki, Y., and Jacobson, M. D., 1993, Programmed cell death and the control of cell survival: Lessons from the nervous system, *Science* **262**:695–700.

Rohrmann, G. F., 1992, Baculovirus structural proteins, *J. Gen. Virol.* **73**:749–761.

Rohrmann, G. F., and Leisy, D. J., 1995, Interactions of baculovirus replication proteins: Application of a yeast two-hybrid system, *American Society for Virology Annual Meeting Abstracts*, p. 105.

Rothe, M., Pan, M.-G., Henzel, W. J., Ayres, T. M., and Goeddel, D. V., 1995, The TNFR2-TRAF signalling complex contains two novel proteins related to baculoviral inhibitor of apoptosis proteins, *Cell* **83**:1243–1252.

Roy, N., Mahadevan, M. S., McLean, M., Shutler, G., Yaraghi, Z., Farahani, R., Baird, S., Besner-Johnston, A., Lefebvre, C., Kang, X., Salih, M., Aubry, A., Tamai, K., Guan, X., Ioannou, P., Crawford, T. O., de Jong, P. J., Surh, L.,

Ikeda, J. E., Korneluk, R. G., and MacKenzie, A., 1995, The gene for neuronal apoptosis inhibitory protein is partially deleted in individuals with spinal muscular atrophy, *Cell* **80**:167–178.

Schwartz, L. M., Smith, S. W., Jones, M. E., and Osborne, B. A., 1993, Do all programmed cell deaths occur via apoptosis?, *Proc. Natl. Acad. Sci. USA* **90**: 980–984.

Steller, H., 1995, Mechanisms and genes of cellular suicide, *Science* **267**:1445–1449.

Strasser, A., Harris, A. W., Jacks, T., and Cory, S., 1994, DNA damage can induce apoptosis in proliferating lymphoid cells via p53-independent mechanisms inhibitable by Bcl-2, *Cell* **79**:329–339.

Sugimoto, A., Friesen, P. D., and Rothman, J. H., 1994, Baculovirus *p35* prevents developmentally programmed cell death and rescues a *ced*-9 mutant in the nematode *Caenorhabditis elegans*, *EMBO J.* **13**:2023–2028.

Thompson, C. B., 1995, Apoptosis in the pathogenesis and treatment of disease, *Science* **267**:1456–1462.

Uren, A. G., Pakusch, M., Hawkins, C. J., Puls, K. L., and Vaux, D. L., 1996, Cloning and expression of apoptosis inhibitory protein homologs that function to inhibit apoptosis and/or bind tumor necrosis factor receptor-associated factors, *Proc. Natl. Acad. Sci. USA* **93**:4974–4978.

Vaux, D. L., and Hacker, G., 1995, Hypothesis: Apoptosis caused by cytotoxins represents a defensive response that evolved to combat intracellular pathogens, *Clin. Exp. Pharm. Physiol.* **22**:861–863.

Vaux, D. L., and Strasser, A., 1996, The molecular biology of apoptosis, *Proc. Natl. Acad. Sci. USA* **93**:2239–2244.

Vaux, D. L., Weissman, I. L., and Kim, S. K., 1992, Prevention of programmed cell death in *Caenorhabditis elegans* by human *bcl*-2, *Science* **258**:1955–1957.

Vaux, D. L., Hacker, G., and Strasser, A., 1994, An evolutionary perspective on apoptosis, *Cell* **76**:777–779.

Walker, N. I., Harmon, B. V., Gobe, G. C., and Kerr, J. F. R., 1988, Patterns of cell death, *Methods Achiev. Exp. Pathol.* **13**:18–54.

White, K., Grether, M. E., Abrams, J. M., Young, L., Farrell, K., and Steller, H., 1994, Genetic control of programmed cell death in *Drosophila*, *Science* **264**: 677–683.

White, K., Tahaoglu, E., and Steller, H., 1996, Cell killing by the *Drosophila* gene reaper, *Science* **271**:805–807.

Williams, G. T., and Smith, C. A., 1993, Molecular regulation of apoptosis: Genetic controls on cell death, *Cell* **74**:777–779.

Wolff, T., and Ready, D. F., 1991, Cell death in normal and rough eye mutants of *Drosophila*, *Development* **113**:825–839.

Xue, D., and Horvitz, H. R., 1995, Inhibition of the *Caenorhabditis elegans* cell-death protease CED-3 by a CED-3 cleavage site in baculovirus p35 protein, *Nature* **377**:248–251.

Yuan, J., and Horvitz, H. R., 1992, The *Caenorhabditis elegans* cell death gene *ced*-4 encodes a novel protein and is expressed during the period of extensive programmed cell death, *Development* **116**:309–320.

Yuan, J., Shaham, S., Ledoux, S., Ellis, H. M., and Horvitz, H. R., 1993, The *C. elegans* cell death gene *ced*-3 encodes a protein similar to mammalian interleukin-1beta-converting enzyme, *Cell* **75**:641–652.

6

Stress Protein Gene Expression in Amphibians

JOHN J. HEIKKILA, ADNAN ALI, NICK OHAN, and YING TAM

1. INTRODUCTION

Prokaryotic and eukaryotic organisms respond at the cellular level to environmental or chemical stressors such as elevated temperature, sodium arsenite, or exposure to heavy metals with the expression of a set of heat shock or stress protein (hsp) genes (reviewed by Nover, 1991; Parsell and Lindquist, 1993; Morimoto *et al.*, 1994). A number of hsp gene family members are also expressed normally within the cell and appear to function as molecular chaperones and are involved in protein folding, assembly, and transport. hsp gene expression has also been correlated with the acquisition of thermotolerance. It is likely that during cellular stress hsps bind to and prevent irreversible aggregation or misfolding of damaged or denatured proteins. Therefore, this class of stress proteins is essential under normal growth conditions as well as serving to protect the cell from the adverse effects of stress.

JOHN J. HEIKKILA, ADNAN ALI, NICK OHAN, and YING TAM • Department of Biology, University of Waterloo, Waterloo, Ontario N2L 3G1, Canada.

Stress-Inducible Processes in Higher Eukaryotic Cells, edited by Koval. Plenum Press, New York, 1997.

These proteins have been classified on the basis of size into three main families, namely, the small hsps (16–36 kDa), the hsp70s (68–73 kDa), and the high-molecular-weight hsp90s (80–110 kDa) (Parsell and Lindquist, 1993; Morimoto *et al.*, 1994). The hsp70 and hsp90 families are extremely well conserved, with genes and proteins from a range of species sharing a very high degree of sequence similarity. In contrast, the small hsps exhibit the lowest degree of conservation of all hsp families. Nevertheless, the small hsps are related based on limited nucleotide and amino acid sequence similarity and conserved protein structure. Ubiquitin genes have also been shown to be induced by heat shock and thus constitute an additional family of stress-inducible genes (Bond and Schlesinger, 1985; Ozkaynak *et al.*, 1987).

The heat-inducible regulation of hsp gene expression occurs primarily at the transcriptional level although regulation at the level of mRNA stability and translation have been documented (reviewed by Nover, 1991; Parsell and Lindquist, 1993; Morimoto *et al.*, 1994). Heat shock-induced transcriptional activation of hsp genes is mediated by the heat shock element (HSE) found in the 5′ upstream region of these genes and interacts with a transcriptional activating protein known as heat shock factor (HSF). HSF is present in normal cells as an inactive monomer that can form an active trimer on heat shock and is then able to bind to the HSE and facilitate transcription of the hsp gene.

Another family of stress-inducible proteins that are related to the hsp70 family are the glucose-regulated proteins (grps) (reviewed in Lee, 1992; Gething *et al.*, 1994). These grps are located within the endoplasmic reticulum and comprise approximately 5% of the luminal protein content. Like hsp70, the grps act as chaperones and have a role in the folding and assembly of newly synthesized protein. In most organisms, there are at least two grps whose sizes fall in the ranges 73–78 and 94–100 kDa and are referred to as grp78 and grp98, respectively. grp78 is also known as the immunoglobulin heavy chain binding protein (BiP). The expression of grp genes in eukaryotic cells is enhanced under conditions of glucose starvation, inhibitors of glycosylation, and sulfhydryl-reducing compounds. Mammalian grp78 promoters have been shown to contain a relatively complex set of *cis*-acting regulatory elements both for basal expression and for stress induction (Lee, 1992; Gething *et al.*, 1994).

Most of the research examining eukaryotic stress protein gene expression has been carried out in *Drosophila*, yeast, and mammalian tissue culture cells. Gradually, additional information has accumulated in other eukaryotic systems. The following discussion reviews the current knowledge regarding expression of stress protein genes in amphibians. This review will detail the isolation and characterization of amphibian stress protein genes as well as examine their expression in tissue culture, embryo and adult systems.

2. ISOLATION AND CHARACTERIZATION OF AMPHIBIAN STRESS PROTEIN GENES

2.1. hsp70 Gene Family

Some of the first amphibian hsp genes isolated and sequenced were four members of the *Xenopus laevis* (African clawed toad) hsp70 gene family, namely, hsp70A, hsp70B, hsp70C, and hsp70D (Table I; Bienz, 1984a; Bienz and Pelham, 1982; Horrell *et al.*, 1987). Introns were not detected in any of the four members of the *Xenopus* hsp70 gene family (Bienz, 1984b). Analysis of a representative gene, hsp70A, revealed that it had 74% identity with *Drosophila* hsp70. The 5' regulatory region contained a TATA and CCAAT box as well as the heat shock consensus element, HSE. A cDNA encoding a heat-inducible member of the hsp70 family from the salamander *Pleurodeles waltl* was also isolated and sequenced (Billoud *et al.*, 1993). At the protein level, *P. waltl* hsp70 displayed 85.9% identity with *Xenopus* hsp70, 83.3% with rat, and 78.3% with *Drosophila*.

Recently, a full-length cDNA encoding a 70-kDa constitutive member of the *Xenopus* hsp70 family, hsc70.I, has been isolated and sequenced (Ali *et al.*, 1996a). At the protein level, hsc70.I has a similarity of over 92% with rat, mouse, and bovine hsc70. The amino acid sequence along the length of the protein including the ATP-binding domain is conserved between rat hsc71 and *Xenopus* hsc70.I. The carboxyl region is the most divergent portion of the protein, i.e., approximately 85% identity. Interestingly, the similarity of *Xenopus* hsc70.I with heat-inducible *Xenopus* hsp70 is lower at 80%. Also, the similarity in the carboxyl region of the

TABLE I
Isolated Amphibian Stress Protein Genes and cDNAs

Xenopus laevis	
hsp90 gene fragment	Ali *et al.* (1996b)
hsp70 cDNA (X16)	Bienz (1984a)
hsp70A, hsp70B, hsp70C, hsp70D genes	Bienz (1984b)
hsc70.I cDNA	Ali *et al.* (1996a)
hsp30 cDNAs (X4, X5)	Bienz (1984a)
hsp30A, hsp30B genes	Bienz (1984b)
hsp30C, hsp30D, hsp30E genes	Krone *et al.* (1992)
ubiquitin cDNA	Dworkin-Rastl *et al.* (1984)
Rana catesbeiana	
hsp30 gene	Helbing *et al.* (1996)
Pleurodeles waltl	
hsp70 cDNA	Billoud *et al.* (1993)
hsp90 cDNA	Coumailleau *et al.* (1995)

proteins is only 58%. A similar finding has been reported for the carboxyl regions of monkey hsc70 and hsp70 (Sainis *et al.*, 1994). It has been suggested that the divergence of hsp70 from hsc70 genes occurred long before the emergence of amphibians in vertebrate evolution (Ali *et al.*, 1996a). The conservation of the hsp70 linkage group in the major histocompatibility complex from amphibians to mammals also supports this contention (Salter-Cid *et al.*, 1994).

A nuclear localization signal (NLS) that has been reported in a variety of hsp70 and hsc70 proteins from different organisms (Dang and Lee,1989; Mandell and Feldherr, 1992; Rensing and Maier, 1994) was also detected in *Xenopus* hsp70, hsc70.I, and *P. waltl* hsp70 amino acid sequences (Bienz, 1984b; Billoud *et al.*, 1993; Ali *et al.*, 1996a). It is probable that this sequence is functional in *Xenopus* as immunolocalization studies revealed that heat shock induced the translocation of both *Xenopus* hsp70 and hsc70 into the nucleus (Herberts *et al.*, 1993). Another conserved feature found in *Xenopus* hsc70.I and hsp70 and *P. waltl* hsp70 is the carboxyl-terminal sequence EEVD. This sequence is found in hsp70 family members in a variety of organisms (Gunther and Walter, 1994). The EEVD motif of human hsp70 appears to be essential for several chaperonin functions including intramolecular coupling of ATP, substrate binding activities, and for intermolecular interactions between hsp70 and the human DnaJ homologue (Freeman *et al.*, 1995). The finding that amphibian members of the hsp70 family contain this EEVD sequence suggests that they may also have a similar mode of chaperonin activity.

2.2. hsp90 Genes

A cDNA encoding a portion of *P. waltl* hsp90 has been isolated and sequenced (Coumailleau *et al.*, 1995). Interestingly, this sequence had a higher homology to the mammalian and avian β-form of hsp90 rather than to the α-form. The β-form of hsp90 was expressed more strongly at normal physiological temperatures (Barnier *et al.*, 1987; Legagneux *et al.*, 1989). Furthermore, *P. waltl* hsp90 contains an EEVD sequence as found in members of the hsp70 family. A PCR-amplified genomic DNA fragment corresponding to the 5′ region of the hsp90 β-like gene has also been isolated in *X. laevis* (Ali *et al.*, 1996b). The deduced *Xenopus* hsp90β amino acid sequence revealed a strong identity with the hsp90β gene from zebrafish (94%) and human (95%) and a slightly lower degree of identity with hsp90α (90–91%) from the same organisms. In *Xenopus* the hsp90 gene appears to code for a protein of approximately 87 kDa (Heikkila *et al.*, 1987; Darasch *et al.*, 1988).

2.3. Small hsp Genes

In *X. laevis*, five hsp30 genes, hsp30A to hsp30E (in two clusters), as well as two hsp30 cDNAs, X4 and X5, have been cloned and sequenced (Bienz, 1984a,b; Krone *et al.*, 1992; Ali *et al.*, 1993). DNA sequence analysis revealed that hsp30A and hsp30B are not representative of the hsp30 gene family as hsp30A contains a 21-bp insertion in the coding region and hsp30B appears to be a pseudogene (Bienz, 1984a,b). A second gene cluster has been isolated that contains two complete hsp30 genes, hsp30C and hsp30D, as well as a portion of a third gene, hsp 30E (Krone *et al.*, 1992). All of these genes are intronless. Comparison of the DNA sequence of the hsp30C gene with the previously published sequence of hsp30A DNA (Bienz, 1984b) revealed a high degree of similarity (97%) between the two. In contrast, the hsp30D gene is only 75% similar to those hsp30 genes. Both the hsp30C and hsp30D genes encode 24-kDa proteins. The upstream regulatory regions of hsp30A and hsp30C genes contain two TATA boxes, three HSEs (one single and two present as an overlapping doublet), and a downstream CCAAT box at the same locations. The 3' end of the hsp30C gene is AT rich (76%) and also contains a polyadenylation signal and the sequences 5'UAUUUA-3', believed to be involved in the regulation of mRNA stability (Brawerman, 1987), and 5'UUUUUAU-3', which is involved in activation of polyadenylation during *Xenopus* oocyte maturation (McGrew *et al.*, 1989). Recently, an hsp30 gene encoding a 25-kDa polypeptide has been isolated from the frog *Rana catesbeiana* (Helbing *et al.*, 1996). Like *Xenopus* hsp30, the *R. catesbeiana* small hsp gene does not contain any introns. The *R. catesbeiana* gene also contains a TATA box as well as one complete heat shock element in the upstream regulatory region. Interestingly, the *Rana* hsp30 and *Xenopus* hsp30C have only 67% identity at the nucleotide level and 51% identity at the amino acid level. However, the percent similarity at the protein level increases to 76% when conservative amino acid changes are taken into account.

2.4. Ubiquitin Genes

In a number of systems including chicken and yeast it has been shown that heat shock can enhance ubiquitin gene expression (Bond and Schlesinger, 1985; Ozkaynak *et al.*, 1987). Furthermore, sequences comprising the HSE have been identified in the 5' promoter region of both chicken and yeast ubiquitin. A cDNA clone for ubiquitin has been isolated from *Xenopus* (Dworkin-Rastl *et al.*, 1984). This cDNA clone was found to contain multiple tandem repeats of a 76-amino-acid coding sequence that completely matched the corresponding region of human ubiquitin. As discussed later in this chapter (Section 4.3.2) heat shock can also induce the accumulation of ubiquitin mRNA in *Xenopus* embryos.

3. EFFECT OF HEAT AND OTHER STRESSORS ON AMPHIBIAN TISSUE CULTURE CELLS

hsp gene expression has been characterized in a kidney epithelial cell line from *X. laevis* as well as in *R. catesbeiana* epidermal and epithelial primary cell cultures (Voellmy and Rungger, 1982; Ketola-Pirie and Atkinson, 1983; Heikkila *et al.*, 1987;

Darasch *et al.*, 1988) For example, elevation of the incubation temperature of *Xenopus* kidney epithelial A6 cells from 20°C to 33–35°C resulted in the enhanced synthesis of a number of hsps including hsp87, hsp70, and hsp30 (Darasch *et al.*, 1988). Two-dimensional PAGE of the A6 cell hsps revealed the presence of 2 forms of hsp87, 5 hsps in the hsp70 family (pI 5.7 to 5.85) of which 2 were constitutive, and 16 different stress-inducible proteins in the hsp30 family (pI 5.3 to 6.0). The complexity of the *Xenopus* small hsp family has not been observed in other animal systems but has been noted in plants (Nover, 1991). It is not known whether all of these proteins represent distinct gene products or whether some of them are the result of posttranslational modification. Recently, an hsp30 antibody has been developed that recognizes 8 members of the *Xenopus* hsp30 family (Tam and Heikkila, 1995). A more detailed discussion of these findings is given later (Section 4.3.4). In amphibian cells, the synthesis of hsps in response to heat appears to be related to enhanced transcription as the mRNAs encoding these proteins also accumulate during stress (Ketola-Pirie and Atkinson, 1983; Heikkila *et al.*, 1987; Darasch *et al.*, 1988). The effect of different heat shock temperatures as well as time course and recovery conditions on the levels of hsp90, hsp70, and hsp30 mRNA accumulation in *Xenopus* A6 cells has also been determined (Darasch *et al.*, 1988; Ali *et al.*, 1996b). In general, these studies show that the accumulation of decay of hsp mRNA levels occur in a coordinate fashion and parallel the level of hsp synthesis.

Stresses other than heat shock have been found to induce hsp gene expression in amphibian cell lines. For example, culture shock, or the cellular trauma associated with the preparation of fresh primary cultures of *Xenopus* liver, lung, and testis can induce the transient synthesis of hsp70 and hsp85 (Wolffe *et al.*, 1984). A comparison of the effect of heat shock and sodium arsenite on hsp gene expression in the *Xenopus* kidney epithelial cell line, A6, was carried out by Darasch *et al.* (1988). In this study, they found that heat shock (35°C) induced the synthesis of hsp30, hsp51, hsp54, hsp62, hsp70, hsp73, and hsp87. However, sodium arsenite (50 mM) additionally induced hsp37, hsp57, hsp62, and hsp100 whereas hsp51 and hsp54 were not enhanced. This finding suggests that the two stressors induced a common set of hsps as well as proteins that were stressor-specific. This study also found that continuous exposure of A6 cells to either heat shock or sodium arsenite induced transient but distinct temporal patterns of hsp gene expression. For example, heat shock-induced hsp synthesis was detectable within 1 hr and peaked by 2–3 hr whereas maximal sodium arsenite-induced hsp synthesis did not occur until 12 hr. The transient nature of the hsp response has been demonstrated in a number of different organisms. Petersen and Lindquist (1989) have shown that regulation of hsp70 mRNA degradation during recovery from heat shock in *Drosophila* is mediated by sequences present in the 3′ noncoding region including the UAUUUA consensus sequence, which has been shown to confer instability to a number of mammalian messages (Brawerman, 1987). It has been proposed that hsp70 mRNA degradation may be mediated by the same mechanism that is responsible for destabilizing these mRNAs and that the mechanism is nonfunctional at heat shock temperatures (Petersen and Lindquist, 1989). Following heat shock, the mechanism would be reactivated and could recognize signals present in the 3′ end of hsp70 mRNA that target it for degradation. As mentioned previously, the 3′ end of the hsp30C gene contains a perfect match to the UAUUUA consensus sequence (Krone *et al.*, 1992). Thus, degradation of hsp mRNAs during recovery from heat shock in *Xenopus* may be regulated in a similar fashion.

A synergistic enhancement of hsp synthesis and hsp mRNA accumulation in A6 cells after treatment with combined heat shock and sodium arsenite has also been observed (Heikkila *et al.*, 1987; Ali *et al.*, 1996b). For example, treatment of A6 cells with 10 mM sodium arsenite and a 30°C heat shock produced a larger increase in hsp synthesis and hsp mRNA accumulation than the sum of the individual stresses. The mechanism involved is not known but may be related to an increase in the rate of transcription or stability of hsp mRNA.

Another stress protein family that has been characterized in *X. laevis* A6 cells is the glucose-regulated proteins. Exposure of A6 cells to 2-deoxyglucose, tunicamycin, 2-deoxygalactose, and dithiothreitol enhanced the synthesis of grp78 and grp98 as well as the accumulation of grp78 mRNA (Winning *et al.*, 1989). In this study, *Xenopus* grp78 cross-reacted with chicken grp78 suggesting that the *Xenopus* protein shares homology with the corresponding avian grp78. Furthermore, these data as well as time course and recovery experiments suggested that A6 cells have a grp response similar, but not identical, to that found in mammalian cells.

4. STRESS PROTEIN GENE EXPRESSION DURING AMPHIBIAN OOGENESIS AND EMBRYOGENESIS

4.1. Constitutive Expression of Stress Protein Gene Expression during Oogenesis

hsp70 and hsp70 mRNA are detectable constitutively in *Xenopus* oocytes at normal incubation temperatures (Bienz, 1982, 1984a; Bienz and Gurden, 1982; Davis and King, 1989; Browder *et al.*, 1987; Horrell *et al.*, 1987). For example, hsp70A and hsp70B mRNA accumulation was detected by RNase protection analysis during early oogenesis reaching maximum levels by approximately stage III (Horrell *et al.*, 1987). Furthermore, these transcripts were retained stably during oocyte maturation, fertilization, and early cleavage stages of embryogenesis. Immunocytochemical and biochemical analyses have indicated that hsp70 was detectable at all stages of *Xenopus* oogenesis and that it was present throughout the oocyte but had relatively higher concentrations in mitochondria and the nucleus (Herberts *et al.*, 1993). Constitutive expression of a somatic heat-inducible hsp70 gene during *P. waltl* oogenesis has also been examined (Billoud *et al.*, 1993). This study found that both hsp70 mRNA and hsp70 increased in oocytes from stage II to VI. This phenomenon appears to be the result of transcriptional activity based on *in situ* hybridization data obtained with lampbrush chromosome loops. Immunolocalization studies of hsp70 during oogenesis determined that hsp70 proteins were localized in the cytoplasm of young oocytes and then transferred to the nucleus during oocyte maturation. This latter finding is supported by previous work examining the recycling of two hsp70-related peptides across the nuclear envelope in *Xenopus* oocytes (Mandell and Feldherr, 1990).

The presence of both hsp90 protein and mRNA throughout oogenesis of *P. waltl* has been demonstrated (Coumailleau *et al.*, 1995). Furthermore, immunolocalization studies determined that hsp90 was found primarily in the cytoplasm in early oocytes whereas in late stage VI *P. waltl* oocyte hsp90 levels were enriched in the nucleus. Similarly, hsp90 mRNA has been detected in *Xenopus* unfertilized eggs, indicating that hsp90 gene expression occurs during oogenesis (Ali *et al.*, 1996b). Finally, the other major family of hsps, the hsp30s, have not been detected constitutively in oocytes or unfertilized eggs, suggesting that this gene family is not expressed during oogenesis (Bienz, 1984a; Krone and Heikkila, 1988; 1989).

Recently, coimmunoprecipitation experiments using a monoclonal anticentrin antibody have detected a complex consisting not only of centrin but also hsp70 and hsp90 in cytostatic factor-arrested *Xenopus* oocytes (Uzawa *et al.*, 1995). Furthermore, centrin is released from this cytoplasmic complex on oocyte activation presumably as a result of an increase in the levels of cytoplasmic free calcium. This suggests that hsp70 and hsp90 may play a role in sequestering centrosome components during early development. The authors also suggest that complex formation with hsp70 and hsp90 may be a general mechanism by which oocyte protein is stored in an inactive state until required by the embryo.

4.2. Effect of Heat Shock on Gene Expression in Oocytes

An early study examining the effect of heat shock on hsp gene expression in *Xenopus* oocytes reported that while heat shock inhibited the normal rate of protein synthesis there was an increase in hsp70 production (Bienz and Gurdon, 1982). Furthermore, this phenomenon occurred with enucleated or α-amantin-injected oocytes, suggesting that the response was controlled at the translational level. Other studies also reported that heat shock induced the synthesis of hsp70 in *Xenopus* and European green frog oocytes (Rojas and Allende, 1983; Baltus and Hanocq-Quertier, 1985; Chen and Stumm-Zollinger, 1986). The view that oocytes have a heat shock response was challenged by King and Davis (1987) and Horrell *et al.* (1987) who were unable to detect the synthesis of hsp70 in heat-shocked oocytes. Both of these studies indicated that contamination of the oocytes with follicular cells was probably responsible for the detection of hsp70 after heat shock in the earlier investigations. In the key experiments by King and Davis (1987) and Horrell *et al.* (1987) establishing the lack of heat shock-induced hsp70 synthesis in *Xenopus* oocytes, either equivalent embryo or protein samples were examined by electrophoresis. Browder *et al.* (1987), who examined the electrophoretic profile of proteins on the basis of equivalent acid-insoluble radio-activity, did find a change in the protein pattern of oocytes and body cavity eggs (which are devoid of follicular cell contamination) after heat shock with the synthesis of a number of stress proteins including hsp70. As protein synthesis was extensively inhibited under their heat shock conditions, this study does not support a true induction of hsp synthesis above control levels but rather reflects a change in the pattern of residual protein synthesis in *Xenopus* oocytes and eggs. The function(s) of this low level of hsp synthesis is not clear.

The *Xenopus* oocyte has also been used as an *in vivo* expression system to monitor the transcriptional efficiency of hsp promoters. Voellmy and Rungger (1982) found that a microinjected *Drosophila* hsp70 gene was heat-inducible in *Xenopus* oocytes. However, Bienz (1984b, 1986) found that whereas a microinjected *Xenopus* hsp30 gene was heat-inducible in oocytes, a *Xenopus* hsp70 gene was expressed constitutively but was not heat-inducible and that this activity was related to the HSE and CCAAT elements. Recently, Landsberger *et al.* (1995; Landsberger and Wolffe, 1995) reexamined this phenomenon and found that if a *Xenopus* hsp70 promoter/chloramphenicol acetyl transferase (CAT) construct was injected into oocyte nuclei and given additional time for assembly into chromatin, then this promoter was heat-inducible. Furthermore, deletion of the CCAAT boxes resulted in a loss of promoter activity. Although these results demonstrate that a microinjected *Xenopus* hsp70 gene becomes heat-inducible in oocytes after it is assembled into chromatin, the question of whether endogenous hsp70 genes are expressed during heat shock was not addressed.

4.3. hsp Gene Expression during Early Amphibian Embryogenesis

4.3.1. Constitutive Expression of hsp Genes during Amphibian Development

hsp90, hsp70, and hsc70 mRNAs and protein are detectable throughout *Xenopus* development including unfertilized and fertilized eggs and cleavage stage embryos (Heikkila *et al.*, 1987; Horrell *et al.*, 1987; Krone and Heikkila, 1989; Davis and King, 1989; Uzawa *et al.*, 1995; Ali *et al.*, 1996a,b). Additionally, hsp70-

related proteins may be preferentially localized in specific cell types during *Xenopus* development as immunocytochemical analysis has detected an increased concentration of hsp70 in the nucleus and perinuclear region of involuted cells of the marginal zone in gastrula embryos (Herberts *et al.*, 1993). Whereas hsc70 mRNA levels increase during *Xenopus* development after the midblastula transition (MBT), the relative levels of hsc70 mRNA are unaffected by heat shock in contrast to hsp70 mRNA (Ali *et al.*, 1996a). Even though the *Xenopus* embryo hsc70 genes were not induced on heat shock, it is possible that these constitutively expressed hsc70 genes may contribute to the thermoresistance of the embryo. In fact, an increase in the enhanced accumulation of hsc70 mRNA coincides with the acquisition of thermoresistance of *Xenopus* embryos after MBT (Heikkila *et al.*, 1985; Ali *et al.*, 1996a). In support of a protective role for hsc70, it has been reported that microinjection of bovine hsc70 into *Xenopus* oocytes reduced the response of a coinjected heat shock reporter plasmid to thermal stress (Mifflin and Cohen, 1994). It is tenable that exposure of embryos to heat shock could lead initially to the interaction of hsc70 protein with denatured and misfolded proteins prior to an increase in effective hsp70 levels.

4.3.2. Heat Shock-Induced hsp Gene Expression in Post-MBT Embryos

Heat shock-induced expression of hsp genes is developmentally regulated during amphibian development. For example, a number of studies have found that hsp70 and/or hsp70 mRNA do not accumulate in heat-shocked, sodium arsenite- or ethanol-treated *Xenopus* embryos until after the midblastula stage of development (Table II; Bienz, 1984a,b; Heikkila *et al.*, 1985, 1987; Nickells and Browder, 1985; Horrell *et al.*, 1987; Browder *et al.*, 1987; Davis and King, 1989; Krone and Heikkila, 1989). Furthermore, the acquisition of heat-inducible hsp gene expression at the midblastula stage coincides with enhanced thermoresistance (Heikkila *et al.*, 1985; Nickells and Browder, 1985). The midblastula stage or MBT is a key stage during early development that is associated with an increase in the duration of the cell cycle, loss of synchronous cell division, a reduction in the rate of DNA synthesis, and the activation of transcription of selected genes from the embryonic genome (Newport and Kirschner, 1982a,b).

TABLE II
Summary of Heat Shock-Induced Accumulation of hsp mRNA during Early Xenopus Development

hsp mRNA	Developmental stages (1 hr heat shock at 33°C)						
	Cleavage	Blastula	Gastrula	Neurula	Early tailbud	Mid-tailbud	Tadpole
hsp90[a,d]	–	+	+	+	+	+	+
hsp70[a,b,d]	–	+	+	+	+	+	+
hsc70[a,d]	–	–	–	–	–	–	–
ubiquitin[a,d]	–	+	+	+	+	+	+
PTB hsp30[a]	–	+	+	+	?	?	?
hsp30A[b]	–	–	–	–	+	+	+
hsp30C[b,c]	–	–	–	–	+	+	+
hsp30D[c]	–	–	–	–	–	+	+

[a] Determined via Northern blotting.
[b] Determined via RNase protection assay.
[c] Determined via RT-PCR.
[d] Constitutive levels detected throughout development.

Other hsp genes that are heat-inducible after the MBT are ubiquitin and hsp90 (Heikkila *et al.*, 1987; Ovsenek and Heikkila, 1988; Ali *et al.*, 1996b). The biological significance of increased levels of ubiquitin in heat-shocked embryos is not known. If heat shock results in an increased synthesis of abnormal or denatured protein in the embryo, increased synthesis of ubiquitin could facilitate its role in ATP-dependent proteolysis as suggested by Ciechanover *et al.* (1984).

The stage-dependent expression of hsp gene expression during early *Xenopus* development has also been observed in other animal systems. For example, during *Drosophila* and sea urchin development, heat-induced hsp synthesis is not observed until embryos reach the blastoderm and blastula stage, respectively (reviewed in Heikkila, 1993a,b). Also, mouse and rabbit embryos are incompetent for heat-induced hsp synthesis during the early cleavage stages of development but do synthesize hsp70 at the blastocyst stage. Given the similarity in the timing of the acquisition of the heat shock response in insect, echinoderm, amphibian, and mammalian development, it is likely that this phenomenon has been conserved through evolution.

hsps in addition to hsp70 and hsp90 are also heat-inducible shortly after the onset of MBT in *Xenopus*. For example, in heat-shocked gastrula embryos, hsps having relative molecular masses of 57, 43, and 35 kDa are detectable (Nickells and Browder, 1985). Interestingly, hsp35 was found to be localized to the vegetal half of the embryo. In a subsequent study, the identity of hsp35 was determined to be glyceraldehyde-3-phosphate dehydrogenase (Nickells and Browder, 1988). It was suggested that heat shock might place an energy demand on *Xenopus* embryos that may be alleviated in part by increasing the levels of specific glycolytic enzymes. This possibility was supported by the finding that hsp62, which is detectable in heat-shocked neurula embryos, was in fact pyruvate kinase (Marsden *et al.*, 1993).

The expression of grp genes has also been examined during early *Xenopus* development. Tunicamycin treatment of *X. laevis* embryos enhanced the synthesis of both grp78 and grp98 (Winning *et al.*, 1991). grp78 mRNA was detectable throughout early development but was not tunicamycin-inducible at stages prior to MBT. A detailed examination of the proteins in the grp78 region in embryo and adult cells by two-dimensional PAGE revealed a total of three peptides. One was observed in both embryos and adult cells, another was adult-specific, and the third protein appeared to be embryo-specific. These results suggested that grp78 synthesis might undergo a switch from an embryonic to an adult pattern during *Xenopus* development. A series of microinjection experiments demonstrated that the regulatory elements associated with the grp78 promoter and mammalian grp78 promoters may be conserved (Winning *et al.*, 1992; Vezina *et al.*, 1994). This is based on the finding that microinjection of promoter deletion mutants of a rat grp78/CAT fusion gene into *Xenopus* embryos required the presence of similar promoter sequences for constitutive and tunicamycin-inducible expression as found in mammalian cells.

4.3.3. Involvement of cis- and trans-acting Factors in the Regulation of Heat-Inducible HSP Gene Expression at the MBT

To examine the mechanisms associated with the developmental stage-dependent expression of hsp70 genes, a chimeric gene containing the *Xenopus* hsp70B promoter fused to a CAT gene was microinjected into fertilized *Xenopus* eggs (Krone and Heikkila, 1989). CAT enzyme assays and RNase protection analysis revealed that heat-induced transcription was activated only after embryos reached the midblastula stage of development. A similar result was obtained by Landsberger *et al.* (1995). These results indicate that the *cis*-acting sequences required for both heat-inducibility and activation of expression after the MBT are present on the *Xenopus* hsp70 promoter.

An important question that was addressed was whether the developmental regulation of hsp70 gene expression was related to HSF availability and/or inducibility (Ovsenek and Heikkila, 1990). A synthetic oligonucleotide corresponding to the proximal HSE of the *Xenopus* hsp70B gene was used in DNA mobility shift assays in an attempt to examine the basis for the stage-dependent expression of hsp genes in *Xenopus* embryos. Interestingly, the presence of HSF binding activity was detected in heat-shocked unfertilized eggs and cleavage stage embryos in which HSP gene expression was inhibited along with normal zygotic transcription. Furthermore, the kinetics of HSF activation and deactivation were similar in both pre- and post-MBT embryos (Karn *et al.*, 1992). A UV-cross-linking technique demonstrated that the relative molecular mass of *Xenopus* HSF is 88 kDa in both cleavage and neurula stages, which is similar to that reported for HSF in HeLa cells (Goldenberg *et al.*, 1988). The finding that both pre-and post MBT forms of HSF are similar in relative molecular mass and in the pattern of activation during time course and recovery experiments, indicates that maternal HSF may be similar or identical to the embryonic form of HSF. Thus, it appears that the mechanism involved in the stage-dependent transcription of hsp genes, such as hsp70 is not related to the absence of activatable HSF binding activity prior to the MBT. However, it is tenable that additional regulatory factors or a posttranslational modification of HSF which may be necessary for transcriptional activation subsequent to HSE binding, are absent or blocked prior to the MBT. Nevertheless, it is clear that the mechanism for the activation of HSF binding is operative throughout early *Xenopus* development, even in the absence of transcriptional activity prior to the MBT.

It is possible that the mechanism associated with the developmental regulation of genes normally activated at MBT also applies to the hsp genes. In the model proposed by Newport and Kirschner (1982a,b) and Kimelman *et al.* (1987), it was suggested that pre-MBT embryos are transcriptionally competent such that they contain functional transcription factors and RNA polymerase II but are unable to synthesize RNA because of the rapid cell cycle. A variety of transcriptional factors have been detected in pre-MBT embryos including those binding to CCAAT, GC, ATF/AP-1, and serum responsive elements (Ovsenek *et al.*, 1990; Mohun *et al.*, 1989). Assuming that HSF found in pre-MBT embryos is functional, it is likely that maternal HSF is involved in a heat shock response at the MBT. The obvious advantage of such a mechanism, given the role of hsps in the acquisition of thermotolerance (Nover, 1991; Parsell and Lindquist, 1993; Morimoto *et al.*, 1994), is that embryos would acquire the ability to transcribe HSP mRNA at the earliest possible time after the MBT.

Recently, a cDNA encoding the *X. laevis* heat shock factor, XHSF1, has been isolated and sequenced (Stump *et al.*, 1995). The HSF cDNA encodes a 451-amino-acid protein which is of comparable size to other vertebrate HSF proteins. Furthermore, at the amino acid level there is similarity in the putative DNA-binding and trimerization domains. *In vitro* synthesized *Xenopus* HSF protein was found to bind specifically with HSE by means of gel shift analysis and DNase I footprinting. Also, the translational product of microinjected XHSF1 mRNA was found to accumulate in the oocyte nucleus and promote the transcription of a subsequently injected *Xenopus* hsp70/CAT construct (Stump *et al.*, 1995; Landsberger and Wolffe, 1995). The availability of the *Xenopus* HSF cDNA should enable researchers to address a number of issues relating to the heat-inducible expression of hsp genes during early development.

4.3.4. Small hsp Gene Expression during Amphibian Development

The members of the small-molecular-weight hsp gene family, hsp30, are differentially expressed in a heat-inducible fashion during early development of *X. laevis* (Table II). Furthermore, the expression of some of the hsp genes is regulated at the level of mRNA stability. Initial studies determined that hsp30A gene expression was not detectable in control embryos and that it was not heat-inducible until the tadpole stage (Bienz, 1984a). However, subsequent RNase protection and reverse-transcription-polymerase chain reaction (RT-PCR) analysis revealed that hsp30A and hsp30C genes were first heat-inducible at the early tailbud stage of development (Krone and Heikkila, 1989; Ali *et al.*, 1993). Also, hsp30D mRNA was not heat-inducible until the mid-tailbud stage of development, which is approximately 1 day later than hsp30A and hsp30C (Ohan and Heikkila, 1995). Heat-inducible hsp30 mRNA accumulation has also been detected in tadpole tissue of *R. catesbeiana* (Helbing *et al.*, 1996). Interestingly, this study also reported the expression of hsp30 genes in liver tissue of non-stressed metamorphosing tadpoles which may be initiated by thyroid hormone. A similar study has not been carried out in *Xenopus*.

The mechanism(s) involved in the differential expression of the small hsps during *Xenopus* development is not known. A possible mechanism for the repression of hsp30 gene expression early in development may be the site-specific methylation of key genomie DNA regions (Adams, 1990). Sequence analysis of the 5'-flanking DNA sequences of the hsp30C gene indicated the presence of potential methylation sites (Krone *et al.*, 1992; Ohan and Heikkila, 1995). However; in a preliminary set of studies, treatment of embyros with 5-azacytidine did not prematurely induce the expression of the hsp30D gene (N. Ohan and J. J. Heikkila, unpublished data). These findings suggest that DNA methylation may not be involved in the regulatory mechanism associated with *Xenopus* small hsp expression.

An additional possibility is that the differential expression of the hsp30 gene family may be controlled at the level of chromatin as differences in chromatin structure have been correlated with differential expression of the small hsp genes in the nematode *Caenorhabditis elegans* (Dixon *et al.*, 1990). Furthermore, as mentioned previously, the *Xenopus* hsp30 genes occur as clusters similar to globin genes, which are regulated at the level of chromatin conformation by a globin locus control region (Lowrey *et al.*, 1992). Furthermore, regulation by chromatin structure would be consistent with previous studies in which it was found that microinjected hsp30A/CAT and hsp30C/CAT constructs were heat-inducible, but prematurely expressed at the midblastula stage (Krone and Heikkila, 1989; Ali *et al.*, 1993). It is tenable that the microinjected constructs did not attain the proper chromatin conformation, probably because of insufficient flanking sequences, and thus were not correctly regulated during development. One could speculate that sequential changes in chromatin structure might result in a sequential expression of the genes within a single cluster. This set of events would account for the later expression of the hsp30D gene as compared with hsp30C.

In *Xenopus*, very low levels of heat-inducible hsp30 mRNA at the late blastula, gastrula, and neurula stages of development have been detected (Ohan and Heikkila, 1995). The relative levels of these transcripts in heat-shocked blastula embryos are approximately 50- to 100-fold less than observed in tailbud embryos. The membership of these pre-tailbud (PTB) mRNAs in the hsp30 family was based on their heat shock-inducible detection by an hsp30 genomic probe and their size. The exact identity and relationship of the PTB hsp30 mRNAs with other members of the hsp30 family will have to await the cloning of their cDNAs. Nevertheless, it is likely that these PTB hsp30 mRNAs are transcribed from some as yet uncharacterized members of the hsp30 gene family. It was also found that treatment of late blastula and gastrula embryos with cycloheximide resulted in an increase in the levels of PTB hsp30 mRNAs after heat shock. Cycloheximide has been shown to stabilize *Eg2* mRNA in *Xenopus* embryos (Duval *et al.*, 1990). The mechanism for cycloheximide-induced stability of unstable mRNAs appears to involve the inhibition of protein synthesis which may reduce the levels of a factor(s) involved in mRNA degradation (Nanbu et al., 1994; Sheu et al., 1994). Regulation of PTB hsp30 gene expression at the level of mRNA stability is supported by the presence of a potential mRNA instability element in the 3' untranslated region of the hsp30C gene (Krone et al., 1992). mRNA stability has also been implicated as one mechanism of regulating the chicken hsp23 gene (Edington and Hightower, 1990). Xenopus PTB hsp30 mRNAs appear to be unique because they are unstable under moderate heat shock conditions that have been shown to stabilize other hsp mRNAs (Heikkila et al., 1987; Krone and Heikkila, 1988). The mechanism(s) associated with the instability of the PTB hsp30 mRNA are not known, but future experiments analyzing the 3' end of these mRNAs may aid in understanding this phenomenon.

Although the work mentioned above provided information regarding the structure and transcription of the Xenopus hsp30 gene family, little was known as to the identity of the resultant proteins. Recently, an antibody prepared against a 15-amino-acid peptide corresponding to the carboxyl end of the putative hsp30C protein sequence was developed to identify members of the hsp30 family in X. laevis embryos (Tam and Heikkila, 1995). Immunoblot analysis was unable to detect any heat-inducible small hsps in cleavage, blastula, gastrula, or neurula stage embryos. However, heat-inducible accumulation of a single protein was first detectable in early tailbud embryos and was still observed at mid-tailbud. At the late tailbud stage an additional 5 hsps were observed whereas a total of 13 small hsps were heat-inducible at the early tadpole stage. Comparison of the pattern of heat-inducible protein recognized by the hsp30C antibody in embryos and cultured A6 cells revealed a number of common and unique proteins. Multiple small hsps have been detected in other organisms such as Drosophila and desert fish (Ingolia and Craig, 1982; White et al., 1994). Also, in C. elegans approximately 13 heat-inducible small hsps have been characterized by two-dimensional PAGE and immunoblot analysis (Hockertz et al., 1991).

The hsp30 peptide antibody does not recognize all of the hsp30 family members in heat-shocked tadpoles or A6 cells probably because of differences in the carboxyl ends of the various hsp30 proteins (Tam and Heikkila, 1995). DNA sequence analysis of the different members of the hsp30 gene family for which the carboxyl-terminal sequence is known suggests that the hsp30 antibody should detect hsp30C, hsp30D, and the protein product of the cDNA clone, X4 (Bienz, 1984b; Krone et al., 1992). To identify the hsp30C protein, a chimeric hsp30C gene, under the control of a constitutive promoter, was microinjected into fertilized eggs and allowed to develop to the gastrula stage, where hsp30 is not normally synthesized, even under heat shock conditions. This series of experiments, in which labeled proteins were isolated and examined by two-dimensional PAGE and immunoblotting, permitted the identification of the hsp30C gene product in early tailbud embryos. This identification was consistent with RT-PCR analysis, which showed that hsp30C mRNA was first heat-inducible at the early tailbud stage of development (Ali et al., 1993). In the future, this novel approach to the identification of the hsp30C gene product may permit the determination of the gene products of other hsp30 genes.

The full spectrum of heat-inducible low-molecular-weight hsps in embryos and somatic cells that reacted to the hsp30C antibody was quite extensive. It is possible that some of these heat-inducible protein spots may be the result of posttranslational modifications such as phosphorylation that has been observed in small hsps in other organisms (Arrigo and Landry, 1994). Although posttranslational modification may be involved in the production of the large hsp30 family in Xenopus, it is possible that multiple genes are also involved as seven different hsp30 genes and cDNAs have been described (Krone et al., 1992).

The sequential heat-induced accumulation of members of the hsp30 family from early tailbud through to tadpole and the difference in the number of heat-inducible proteins recognized by the hsp30C antibody in A6 cells compared with heat-shocked tadpoles support the contention that hsp30 genes are differentially regulated during development (Krone and Heikkila, 1989; Krone et al., 1992; Ali et al., 1993; Ohan and Heikkila, 1995). In Drosophila, the small hsps display a very complex temporal and spatial pattern of gene expression during development (Marin et al., 1993; Arrigo and Landry, 1994). For example, it has been shown that during development hsp27 was localized to the central nervous system, gonads, and imaginal discs at the pupal stage whereas hsp23 was constitutively expressed in the imaginal wing discs (Cheney and Shearn, 1983; Arrigo and Landry, 1994). The differential pattern of hsp30 accumulation during Xenopus development suggests that these proteins may have distinct functions at specific stages of embryogenesis during a stress response. However, it is quite likely that the small hsps act as chaperones within the cell. Furthermore, it has been shown that overexpression of a small hsp gene in mammalian cells can confer thermotolerance (Arrigo and Landry, 1994).

5. hsp GENE EXPRESSION IN AMPHIBIAN ADULT SOMATIC CELLS

Relatively few studies have examined the expression of hsp genes in amphibian adult somatic cells in response to hyperthermia. Bienz (1984a) found that all adult tissues examined displayed heat-induced expression of hsp70 and hsp30 genes but that the levels of hsp30 mRNA varied considerably between cell types. For example, whereas somatic cells each contained approximately $1-3 \times 10^8$ copies of heat-inducible hsp70 mRNA, the levels of hsp30 mRNA varied dramatically with the highest levels in kidney and lower levels in lung, liver, and gut. Constitutive levels of hsp70 as well as the heat shock-induced synthesis of hsp70 and hsp30 and the accumulation of hsp70 mRNA in various tissues of adult salamanders have been reported (Rutledge *et al.*, 1987; Easton *et al.*, 1987; Near *et al.*, 1990; Yu *et al.*, 1994). Interestingly, the accumulation of hsp70 mRNA was also detected in salamanders collected in the field on a hot summer day (Near *et al.*, 1990). In *Xenopus* adults, the relative levels of hsc70 mRNA were found to be different in selected tissues (Ali *et al.*, 1996b). hsc70 mRNA levels were relatively high in spleen and testis with slightly lower levels in eyes, heart, liver, and brain. Relatively low levels of hsc70 mRNA were found in muscle. The reason for the tissue differences in hsc70 mRNA levels is unclear. Aside from the common chaperone-type duties of hsc70 present in all cell types, it is possible that the differences in the relative levels of hsc70 mRNA in various tissues may reflect their specialized cellular environment.

6. hsp GENE EXPRESSION DURING AMPHIBIAN REGENERATION

hsp gene expression has also been examined in regenerating amphibian systems. For example, the trauma associated with the limb amputation in newts has been shown to induce the synthesis of hsps near the wound site within a few hours after injury (Carlone and Fraser, 1989; Tam *et al.*, 1992). Also, Carlone *et al.* (1993) found that injection of retinoic acid at a dose sufficient to initiate proximal–distal reduplication of amputated forelimbs in 50% of treated newts resulted in the enhanced synthesis of a 73-kDa hsp which was maximal 6 days after injection. Finally, in regenerating adult frog sciatic sensory axons it was reported that hsp70 among other proteins produced by nonneuronal cells present in the damaged area were taken up by the injured sensory axon and retrogradely transported to the nerve cell body (Edbladh *et al.*, 1994). It is likely that the enhanced synthesis of hsps is produced in response to the initial trauma as has been demonstrated previously in other systems (Currie and White, 1981; Heikkila and Schultz, 1984). The function of hsps in all of these situations is likely that of a chaperone although other functions related to more specific processes associated with regeneration cannot be ruled out.

7. CONCLUSIONS

It is evident from the preceding studies that hsps and grps have important roles in normal amphibian cells as well as in cells subject to environmental or chemical stress. These stress proteins act as molecular chaperones and are involved in protein transport, folding, and assembly. They may also be involved in sequestering proteins during early amphibian development and maintaining them in an inactive state. Amphibians, especially during the aquatic phase of their life cycle, are particularly susceptible to environmental and chemical stressors. In this context, hsps have been proposed as molecular biomarkers for toxicity, a relationship between the extent of hsp gene expression and the amount of damage inflicted on cells having been demonstrated (Ryan and Hightower, 1994). During stressful conditions it is likely that the chaperonin function of the stress proteins is utilized in response to damaged protein. Clearly additional research on the expression of hsp and grp genes and their function in amphibian systems will have a great impact on our understanding of the heat shock response in general and the functioning of the amphibian cell in particular.

REFERENCES

Adams, R., 1990, DNA methylation, the effect of minor bases on DNA–protein interactions. *Biochem. J.* **265**:309–320.

Ali, A., Krone, P., and Heikkila, J. J., 1993, Expression of endogenous and microinjected *hsp30* genes in early *Xenopus laevis* embryos, *Dev. Genet.* **14**:42–50.

Ali, A., Salter-Cid, L., Flajnik, M., and Heikkila, J. J., 1996a, Isolation and characterization of a cDNA encoding a *Xenopus* 70 kDa heat shock cognate protein, Hsc70.I, *Comp. Biochem. Physiol.* **113B**:681–687.

Ali, A., Krone, P. H., Pearson, D. S., and Heikkila, J. J., 1996b, Evaluation of stress-inducible hsp90 gene expression as a potential molecular biomarker in *Xenopus laevis*, *Cell Stress Chaperones* **1**:62–69.

Arrigo, A.-P. and Landry, J., 1994, Expression and function of the low-molecular-weight heat shock proteins, in: *The Biology of Heat Shock Proteins and Molecular Chaperones* (R. I. Morimoto, A. Tissieres, and C. Georgopoulos, eds.), Cold Spring Harbor Laboratory Press, Cold Spring Harbor, NY, pp. 335–373.

Baltus, E., and Hanocq-Quertier, J., 1985, Heat shock response in *Xenopus* oocytes during meiotic maturation and activation, *Cell Differ.* **16**:161–168.

Barnier, J. V., Bensaude, O., Morange, M., and Babinet, C., 1987, Mouse 89 kDa heat shock protein: Two polypeptides with distinct developmental regulation, *Exp. Cell Res.* **170**:186–194.

Bienz, M., 1982, Expression of *Drosophila* heat shock protein in *Xenopus* oocytes: Conserved and divergent regulatory signals. *EMBO J.* **1**:1583–1588.

Bienz, M., 1984a, Developmental control of the heat shock response in *Xenopus*, *Proc. Natl. Acad. Sci. USA* **81**:3138–3142.

Bienz, M., 1984b, *Xenopus* hsp70 genes are constitutively expressed in injected oocytes, *EMBO J.* **3**:2477–2483.

Bienz, M., 1986, A CCAAT box confers cell-type-specific regulation on the *Xenopus* hsp 70 gene in oocytes, *Cell* **46**:1037–1042.

Bienz, M., and Gurdon, J.B., 1982, The heat shock response in *Xenopus* oocytes is controlled at the translational level, *Cell* **29**:811–819.

Bienz, M., and Pelham, H. R. B., 1982, Expression of a *Drosophila* heat-shock protein in *Xenopus* oocytes: conserved and divergent regulatory signals. *EMBO J.* **1**:1583–1588.

Billoud, B., Rodriguez-Martin, M.-L., Beerard, L., Moreau, N., and Angelier, N., 1993, Constitutive expression of a somatic heat-inducible hsp70 gene during amphibian oogenesis, *Development* **119**:921–932.

Bond, U., and Schlesinger, M. J., 1985, Ubiquitin is a heat shock protein in chicken embryo fibroblasts, *Mol. Cell Biol.* **5**:949–956.

Brawerman, G., 1987, Determinants of messenger RNA stability, *Cell* **48**:5–6.

Browder, L. W., Pollock, M., Heikkila, J. J., Wilkes, J., Wang, T., Krone, P., Ovsenek, N., and Kloc, M., 1987, Decay of the oocyte-type heat shock response of *Xenopus laevis*, *Dev. Biol.* **124**:191–199.

Carlone, R. L., and Fraser, G. A. D., 1989, An examination of heat shock and trauma-induced proteins in the regenerating forelimb of the newt, *Notophthalmus viridescens*, in: *Recent Trends in Regeneration Research* (V. Kiotsis, S. Kovssoulakkos, and H. Wallace, eds.), Plenum Press, New York, pp. 17–25.

Carlone, R. L., Boulianne, R. P., Vijh, K. M., Karn, H., and Fraser, G. A. D., 1993, Retinoic acid stimulates the synthesis of a novel heat shock protein in the regenerating forelimb of the newt, *Biochem. Cell Biol.* **71:**43–50.

Chen, P. S., and Stumm-Zollinger, E., 1986, Patterns of protein synthesis in oocytes and early embryos of *Rana esculenta* complex, Wilhelm, *Roux Arch. Dev. Biol.* **195:**1–9.

Cheney, C. M., and Shearn, A., 1983, Developmental regulation of *Drosophila* imaginal disc proteins: Synthesis of a heat shock protein under non-heat-shock conditions, *Dev. Biol.* **95:**325–330.

Ciechanover, A., Finley, D., and Varshavski, A., 1984, The ubiquitin mediated proteolytic pathway and mechanisms of energy-dependent intracellular protein degradation, *J. Cell Biochem.* **24:**27–53.

Coumailleau, P., Billoud, B., Sourrouille, P., Moreau, N., and Angelier, N., 1995, Evidence for a 90 kDa heat-shock protein gene expression in the amphibian oocyte, *Dev. Biol.* **168:**247–258.

Currie, R. W., and White, F. P., 1981, Trauma-induced protein in rat tissues: A physiological role for a heat shock protein, *Science* **214:**72–73.

Dang, C. V., and Lee, W. M. F., 1989, Nuclear and nucleolar targeting sequences of c-*erb*-A, c-*myb*, N-*myc*, p53, HSP70, and HIV tat proteins, *J. Biol. Chem.* **264:** 18019–18023.

Darasch, S., Mosser, D. D., Bols, N. C., and Heikkila, J. J., 1988, Heat shock gene expression in *Xenopus laevis* A6 cells in response to heat shock and sodium arsenite treatments, *Biochem. Cell Biol.* **66:**862–868.

Davis, R. E., and King, M. L., 1989, The developmental expression of the heat-shock response in *Xenopus laevis*, *Development* **105:**213–222.

Dixon, D. K., Jones, D., and Candido, E. P., 1990, The differentially expressed 16-kD heat shock genes of *Caenorhabditis elegans* exhibit differential changes in chromatin structure during heat shock, *DNA Cell Biol.* **9:**177–191.

Duval, C., Bouvet, P., Omilli, F., Roghi, C., Dorel, C., LeGuellec, R., Paris, J., and Osborne, H. B., 1990, Stability of maternal mRNA in *Xenopus* embryos: Role of transcription and translation, *Mol. Cell Biol.* **10:**4123–4129.

Dworkin-Rastl, E., Shrutkowski, A., and Dworkin, M. B., 1984, Multiple ubiquitin mRNAs during *Xenopus laevis* development contain tandem repeats of 76 amino acid coding sequence, *Cell* **39:**321–325.

Easton, D. P., Rutledge, P. S., and Spotila, J. R., 1987, Heat shock protein induction and induced thermal tolerance are independent in adult salamanders, *J. Exp. Zool.* **241:**263–267.

Edbladh, M., Ekstrom, P. A. R., and Edstrom, A., 1994, Retrograde axonal transport of locally synthesized proteins, e.g., actin and heat shock protein 70, in regenerating adult frog sciatic sensory axons, *J. Neurosci. Res.* **38:**424–432.

Edington, B. V., and Hightower, L. E., 1990, Induction of a chicken small heat shock (stress) protein: Evidence of multilevel posttranscriptional regulation, *Mol. Cell Biol.* **10:**4886–4898.

Freeman, B. C., Myers, M. P., Schumacher, R., and Morimoto, R. I., 1995, Identification of a regulatory motif in Hsp70 that affects ATPase activity, substrate binding and interaction with HDJ-1, *EMBO J.* **14:**2281–2292.

Gething, M.-J., Blond-Egluindi, S., Mori, K., and Sambrook, J. F., 1994, Structure, function, and regulation of the endoplasmic reticulum chaperone, BiP, in: *The Biology of Heat Shock Proteins and Molecular Chaperones* (R. I. Morimoto, A.

Tissieres, and C. Georgopoulos, eds.), Cold Spring Harbor Laboratory Press, Cold Spring Harbor, NY, pp. 111–135.

Goldenberg, C. J., Luo, Y., Fenna, M., Baler, R., Weinmann, R., and Voellmy, R., 1988, Purified human factor activates heat shock promoter in a HeLa cell-free transcription system, *J. Biol. Chem.* **263:**19734–19739.

Gunther, E., and Walter, L., 1994, Genetic aspects of the hsp70 multigene family in vertebrates, *Experientia* **50:**987–1001.

Heikkila, J. J., 1993a, Heat shock gene expression and development. I. An overview of fungal, plant, and poikilothermic animal developmental systems, *Dev. Genet.* **14:**1–5.

Heikkila, J. J., 1993b, Heat shock gene expression and development. II. An overview of mammalian and avian developmental systems, *Dev. Genet.* **14:**87–91.

Heikkila, J. J., and Schultz, G. A., 1984, Different environmental stresses can activate the expression of a heat shock gene in rabbit blastocyst, *Gamete Res.* **10:**45–56.

Heikkila, J. J., Kloc, M., Bury, J., Schultz, G. A., and Browder, L. W., 1985, Acquisition of the heat shock response and thermotolerance during early development of *Xenopus laevis, Dev. Biol.* **107:**483–489.

Heikkila, J. J., Darasch, S. P., Mosser, D. D., and Bols, N. C., 1986, Heat and sodium arsenite act synergistically on the induction of heat shock gene expression in *Xenopus laevis* A6 cells, *Biochem. Cell Biol.* **65:**310–316.

Heikkila, J. J., Ovsenek, N., and Krone, P. H., 1987, Examination of heat shock protein mRNA accumulation in early *Xenopus laevis* embryos, *Biochem. Cell Biol.* **65:**87–94.

Helbing, C., Gallimore, C., and Atkinson, B. G., 1996, Characterization of a *Rana catesbeiana* hsp30 gene and its expression in the liver of this amphibian during both spontaneous and thyroid hormone-induced metamorphosis, *Dev. Genet.* **18:**223–233.

Herberts, C., Moreau, N., and Angelier, N., 1993, Immunolocalization of hsp70-related proteins constitutively expressed during *Xenopus laevis* oogenesis and development, *Int. J. Dev. Biol.* **37:**397–406.

Hockertz, M. K., Clark-Lewis, I., and Candido, E. P. M., 1991, Studies of the small heat shock proteins of *Caenorhabditis elegans* using anti-peptide antibodies, *FEBS Lett.* **280:**375–378.

Horrell, A., Shuttleworth, J., and Colman, A., 1987, Transcript levels and translational control of hsp 70 synthesis in *Xenopus* oocytes, *Genes Dev.* **1:**433–444.

Ingolia, T. D., and Craig, E. A., 1982, Four small *Drosophila* heat shock proteins are related to each other and to mammalian α-crystallin, *Proc. Natl. Acad. Sci. USA* **79:**2360–2364.

Karn, H., Ovsenek, N., and Heikkila, J. J., 1992, Properties of heat shock transcription factor in *Xenopus* embryos, *Biochem. Cell Biol.* **70:**1006–1013.

Ketola-Pirie, C. A., and Atkinson, B. G., 1983, Cold- and heat-shock induction of new gene expression in cultured amphibian cells, *Can. J. Biochem. Cell Biol.* **61:**462–471.

Kimelman, D., Kirschner, M., and Scherson, T., 1987, The events of the mid-blastula transition in *Xenopus* are regulated by changes in the cell cycle, *Cell* **48:**399–407.

King, M. L., and Davis, R. E., 1987, Do *Xenopus* oocytes have a heat shock response? *Dev. Biol.* **119:**532–539.

Krone, P. H., and Heikkila, J. J., 1988, Analysis of hsp 30, hsp 70, and ubiquitin gene expression in Xenopus laevis tadpoles, Development 103:59–67.

Krone, P. H., and Heikkila, J. J., 1989, Expression of microinjected hsp70/CAT and hsp30/CAT chimeric genes in developing Xenopus laevis embryos, Development 106:271–281.

Krone, P. H., Snow, A., Ali, A., Pasternak, J. J., and Heikkila, J. J., 1992, Comparison of the regulatory regions of the Xenopus laevis small heat-shock protein encoding gene family, Gene 110:159–166.

Landsberger, N., and Wolffe, A. P., 1995, Role of chromatin and Xenopus laevis heat shock transcription factor in regulation of transcription from the X. laevis hsp70 promoter in vivo, Mol. Cell. Biol. 15:6013–6024.

Landsberger, N., Ranjan, M., Almouzni, G., Stump, D., and Wolffe, A. P., 1995, The heat shock response in Xenopus oocytes, embryos and somatic cells: A regulatory role for chromatin, Dev. Biol. 170:62–74.

Lee, A. S., 1992, Mammalian stress response: Induction of the glucose-regulated protein, Curr. Opin. Cell Biol. 4:267–273.

Legagneux, V., Mezger, V., Quelard, C., Barnier, J. V., Bensaude, O., and Morange, M., 1989, High constitutive transcription of hsp86 gene in murine carcinoma cells, Differentiation 41:42–48.

Lowrey, C. H., Bodine, D. M., and Nienhuis, A. W., 1992, Mechanisms of DNase I hypersensitive site formation within the human globin locus control region, Proc. Natl. Acad. Sci. USA 89:1143–1147.

Mandell, R. B., and Feldherr, C. M., 1992, The effect of carboxyl-terminal deletions on the nuclear transport rate of rat hsc70, Exp. Cell Res. 198:164–169.

Marin, R., Valet, J. P., and Tanguay, R. M., 1993, Hsp 23 and hsp 26 exhibit distinct spatial and temporal patterns of constitutive expression in Drosophila adults, Dev. Genet. 14:69–77.

Marsden, M., Nickells, R. W., Kapoor, M., and Browder, L. W., 1993, The induction of pyruvate kinase synthesis by heat shock in Xenopus laevis embryos, Dev. Genet. 14:51–57.

McGrew, L. L., Dworkin-Rastl, E., Dworkin, M., and Richter, J. D., 1989, Poly (A) elongation during Xenopus oocyte maturation is required for translational recruitment and is mediated by a short sequence element, Genes Dev. 3: 803–815.

Mifflin, L. C., and Cohen, R. E., 1994, Hsc70 moderates the heat shock (stress) response in Xenopus laevis oocytes and binds to denatured protein inducers, J. Biol. Chem. 269:15718–15723.

Mohun, T. J., Garret, N., and Gurdon, J., 1989, Temporal and tissue-specific expression of the proto-oncogene c-fos during development in Xenopus laevis, Development 107:835–846.

Morimoto, R. I., Tissieres, A., and Georgopoulos, C. (eds.), 1994, The Biology of Heat Shock Proteins and Molecular Chaperones, Cold Spring Harbor Laboratory Press, Cold Spring Harbor, NY.

Nanbu, R., Menoud, P.-A., and Nagamine, Y., 1994, Multiple instability-regulating sites in the 3' untranslated region of the urokinase-type plasminogen activator mRNA, Mol. Cell Biol. 14:4920–4928.

Near, J. C., Easton, D. P., Rutledge, P. S., Dickinson, D. P., and Spotila, J. R., 1990, Heat shock protein 70 gene expression in intact salamanders (Eurycea bislineata) in response to calibrated heat shocks and to high temperatures encountered in the field, J. Exp. Zool. 256:303–314.

Newport, J., and Kirschner, M., 1982a, A major developmental transition in early *Xenopus* embryos: I. Characterization and timing of cellular changes at the midblastula stage, *Cell* **30**:675–686.

Newport, J., and Kirschner, M., 1982b, A major developmental transition in early *Xenopus* embryos: II. Control of the onset of transcription, *Cell* **30**:687–696.

Nickells, R. W., and Browder, L. W., 1985, Region-specific heat-shock protein synthesis correlates with a biphasic acquisition of thermotolerance in *Xenopus laevis* embryos, *Dev. Biol.* **112**:391–395.

Nickells, R. W., and Browder, L. W., 1988, A role for glyceraldehyde-3-phosphate dehydrogenase in the development of thermotolerance in *Xenopus laevis* embryos, *J. Cell Biol.* **107**:1901–1909.

Nover, L., 1991, *The Heat Shock Response*, CRC Press, Boca Raton, FL.

Ohan, N. W., and Heikkila, J. J., 1995, Involvement of differential gene expression and messenger RNA stability in the developmental regulation of the Hsp30 gene family in heat shocked *Xenopus laevis* embryos, *Dev. Genet.* **17**:176–184.

Ovsenek, N., and Heikkila, J. J., 1988, Heat shock induced accumulation of ubiquitin mRNA in *Xenopus laevis* is developmentally regulated, *Dev. Biol.* **12**:582–585.

Ovsenek, N., and Heikkila, J. J., 1990, DNA sequence-specific binding activity of the heat-shock transcription factor is heat inducible before the midblastula transition of early *Xenopus* development, *Development* **110**:427–433.

Ovsenek, N., Williams, G. T., Morimoto, R. I., and Heikkila, J. J., 1990, Cis-acting sequences and trans-acting factors required for constitutive expression of a microinjected HSP70 gene after the midblastula transition of *Xenopus laevis* embryogenesis, *Dev. Genet.* **11**:97–109.

Ozkaynak, E., Finley, D., Solomon, M. J., and Varshavsky, A., 1987, The yeast ubiquitin genes: A family of natural gene fusions, *EMBO J.* **6**:1429–1439.

Parsell, D. A., and Lindquist, S., 1993, The function of heat shock proteins in stress tolerance: Degradation and reactivation of the damaged proteins, *Annu. Rev. Genet.* **27**:437–496.

Petersen, R. B., and Lindquist, S., 1989, Regulation of hsp 70 synthesis by messenger RNA degradation, *Cell Regul.* **1**:135–149.

Rensing, S. A., and Maier, U. G., 1994, Phylogenetic analysis of the stress-70 protein family, *J. Mol. Evol.* **39**:80–86.

Rojas, C., and Allende, J. E., 1983, Amphibian oocytes respond to heat shock after the induction of meiotic maturation by hormones, *Biochem. Int.* **6**:517–525.

Rutledge, P. S., Easton, D. P., and Spotila, J. R., 1987, Heat shock proteins from the lungless salamanders *Eurycea bislineata* and *Desmognathus ochrophaeus*, *Comp. Biochem. Physiol.* **88B**:13–18.

Ryan, J. A., and Hightower, L. E., 1994, Evaluation of heavy-metal ion toxicity in fish cells using a combined stress protein and cytotoxicity assay, *Environ. Toxicol. Chem.* **13**:1231–1240.

Saines, I., Angelidis, C., Pagoulatos, G., and Lazaridis, I., 1994, The hsc70 gene which is slightly induced by heat is the main virus inducible member of the hsp70 gene family. *FEBS Letters* **355**:282–286.

Salter-Cid, L., Kasahara, M. and Flajnik, M. F., 1994, Hsp70 genes are linked to the *Xenopus* major histocompatibility complex, *Immunogenetics* **39**:1–7.

Sheu, J.-J., Jan, S.-P., Lee, H.-T., and Yu, S.-M., 1994, Control of transcription and mRNA turnover as mechanisms of metabolic repression of alpha-amylase gene expression, *Plant J.* **5**:655–664.

Stump, D. G., Landsberger, N., and Wolffe, A. P., 1995, The cDNA encoding *Xenopus laevis* heat-shock factor 1 (XHSF1): Nucleotide and deduced amino-acid sequences, and properties of the encoded protein, *Gene* **160**:207–211.

Tam, Y., and Heikkila, J. J., 1995, Identification of members of the hsp30 small heat shock protein family and characterization of their developmental regulation in heat-shocked *Xenopus laevis* embryos, *Dev. Genet.* **17**:331–339.

Tam, Y., Vethamany-Globus, S., and Globus, M., 1992, Limb amputation and heat shock induce changes in protein expression in the newt, *Notophthalmus viridescens*, *J. Exp. Zool.* **264**:64–74.

Uzawa, M., Grams, J., Madden, B., Toft, D., and Salisbury, J. L., 1995, Identification of a complex between heat shock proteins in CSF-arrested *Xenopus* oocytes and dissociation of the complex following oocyte activation, *Dev. Biol.* **171**:51–59.

Vezina, C., Wooden, S. K., Lee, A. S., and Heikkila, J. J., 1994, Constitutive expression of a microinjected glucose-regulated protein (grp78) fusion gene during early *Xenopus laevis* development, *Differentiation* **57**:171–177.

Voellmy, R., and Rungger, D., 1982, Transcription of a *Drosophila* heat shock gene is heat-induced in *Xenopus* oocytes, *Proc. Natl. Acad. Sci. USA* **79**:1776–1780.

White, C. N., Hightower, L. E., and Schultz, R. J., 1994, Variation in heat-shock proteins among species of desert fishes (Poeciliidae, *Poeciliopsis*), *Mol. Biol. Evol.* **11**:106–119.

Winning, R. S., Heikkila, J. J., and Bols, N. C., 1989, Induction of glucose-regulated proteins in *Xenopus laevis* A6 cells, *J. Cell. Physiol.* **140**:239–245.

Winning, R. S., Bols, N. C., and Heikkila, J. J., 1991, Tunicamycin-inducible polypeptide synthesis during *Xenopus laevis* embryogenesis, *Differentiation* **46**:167–172.

Winning, R. S., Bols, N. C., Wooden, S. K., Lee, A. S., and Heikkila, J. J., 1992, Analysis of the expression of a glucose-regulated protein (GRP78) promoter/CAT fusion gene during early *Xenopus laevis* development, *Differentiation* **49**:1–6.

Wolffe, A. P., Glover, J. F., and Tata, J. R., 1984, Culture shock: Synthesis of heat-shock-like proteins in fresh primary cell cultures, *Exp. Cell Res.* **154**:581–590.

Yu, Z., Magee, W. E., and Spotila, J. R., 1994, Monoclonal antibody ELISA test indicates that large amounts of constitutive hsp-70 are present in salamanders, turtle and fish, *J. Therm. Biol.* **19**:41–53.

7

Heat Stress in Avian Cells

MILTON J. SCHLESINGER

1. INTRODUCTION

Among the many diverse biological materials that have been utilized to study the stress response, avian cells grown in tissue culture have proved to be particularly advantageous. Monolayers of homogeneous populations of 10^6 to 10^8 fibroblasts can be prepared readily from 11-day chicken embryos and cultured for several days in relatively simple media. A stress agent, either chemical or physical, can be applied under controlled conditions for varying lengths of time and the effects on cellular morphology or metabolism easily monitored. In the specific case of a heat shock, a primary culture of chicken embryo fibroblasts "senses" the stress within minutes of a shift up in temperature that corresponds to as little as a 10% increase above the physiological temperature of the bird, which is usually around 41°C. For cells from human tissue, a temperature of 41°C is sufficient to trigger the cellular stress response (Ron and Birkenfeld, 1987). In the avian fibroblast tissue culture system, response to hyperthermic stress—as measured

MILTON J. SCHLESINGER • Department of Molecular Microbiology, Washington University School of Medicine, St. Louis, Missouri 63110.

Stress-Inducible Processes in Higher Eukaryotic Cells, edited by Koval. Plenum Press, New York, 1997.

by enhanced transcription of heat shock genes (i.e., detection of higher levels of mRNAs) and appearance of newly synthesized heat shock proteins—is detected at 42 to 43°C with the full complement of heat shock proteins induced after 30 min at 45°C (Kelley and Schlesinger, 1978). The latter temperature is considered physiological in that the internal temperature of the adult bird reaches 45°C when the bird goes into flight.

Other normal cells of avian origin respond to mild hyperthermic stress; in particular, chicken reticulocytes and lymphocytes change their pattern of protein synthesis to yield enhanced levels of heat shock proteins and lowered amounts of normal protein. With these particular cells, the reticulocytes show only one major heat shock protein and significant decreases in globin synthesis whereas the lymphocytes synthesize two additional heat shock proteins (Morimoto and Fodor, 1984). Cultures of myoblasts and myotubes respond also to thermal stress with the characteristic pattern of new heat shock protein synthesis, but the more differentiated cells require more intense periods of stress and have a different pattern of heat shock proteins (Atkinson, 1981).

Studies of lens from 11-day chicken embryos show a robust response to a 45°C heat shock with activation of heat shock genes and dramatic increases in the synthesis of the three common heat shock proteins (HSP), HSP24, 70, and 90, while normal protein synthesis is maintained (Collier and Schlesinger, 1986a). Intact 18-day chicken embryos subjected to hyperthermia also synthesize heat shock proteins, although as noted with the hemopoietic cells above, organs respond differently (Voellmy and Bromley, 1982). The conventional heat shock protein pattern is detected in red blood cells taken from an intact anemic adult quail subjected to hyperthermia of 42–43°C (Atkinson and Dean, 1985) and in macrophages recovered from 6-week-old white leghorn female chickens treated to raise their core body temperature to 44°C (Miller and Qureshi, 1992). In the latter studies, exposure of the chicken macrophage cell line, MQ-NCSU, to lead acetate leads to formation of the same pattern of heat shock proteins.

In this chapter, several kinds of biological activities that are affected by a hyperthermic stress of avian tissue culture cells are summarized. Many of these changes are found also when these cells have been exposed to relatively alkaline conditions, e.g., a pH of 8.7, or treated with other stress agents, including oxidants (arsenite), heavy metals (cadmium and zinc), organic chemicals (alcohol), or analogues of amino acids, which replace the normal amino acids in polypeptides and lead to abnormal and denatured proteins in the cell. Most of these stressors have been employed to study the stress response in other biological systems, most notably established cell lines derived from mammalian sources and in unicellular organisms such as yeast and bacteria. With few exceptions, the observations reported for the avian cell culture systems are very much like those found in other biological tissue and cells. And, importantly, the avian cell proteins that function as gene activators and as the classical heat shock proteins themselves are very similar in structure to those with similar functions in cells of other biological species (Kelley and Schlesinger, 1982; Morimoto, 1993).

The very high degree of conservation of gene structure and protein function among the highly diverse organisms in the biosphere indicates that the stress response evolved very early in organism development and, thus, was clearly highly advantageous to preservation of the organism. But, paradoxically, it took almost 15 years between the report that the major cellular response to stress in embryos of *Drosophila* was an activation of a small number of genes and the experimental evidence that this response is common to all organisms (reviewed in Schlesinger *et al.*, 1982). In fact, the avian cell system was among the first to show that what was true for a heat stress in *Drosophila* was true also for man (Kelley and Schlesinger, 1978).

The major emphasis during the past 25 years of research on the stress response in biological systems has focused on gene regulation and the function of heat shock proteins. But there are a variety of other changes that occur in cells following stress and their role is not well understood. Furthermore, in evaluating many of these changes it has become important to consider the consequences of the type and extent of a particular stress imposed on the specific biological system under study. For example, strong perturbations of many cellular structures and metabolic activities have been reported following stress conditions so severe that most of the biological material, i.e., cells or tissue or organism, are *irreversibly* damaged. In such situations, the reported changes become complicated by those that normally accompany necrosis and death; thus, they are unlikely to be relevant to the homeostatic situation for which the stress response is considered to have evolved.

2. EARLY EVENTS IN HEAT-STRESSED AVIAN FIBROBLASTS

In the avian tissue culture system studied in my laboratory, we have been particularly sensitive to the potential complication of necrosis and adjusted the stress conditions to those allowing for virtually complete recovery of the cells once the stress has been removed. The measurement we have used to ensure that cells are not irreversibly damaged by the stress treatment is the capacity of the stressed cell to continue protein synthesis at levels close to those of the nonstressed cell. For primary cultures of chicken embryo fibroblasts prepared from 11-day embryos, incubation of cells at 45°C can be done for up to 4 hr before any significant drop occurs in protein synthetic capacity. Within minutes of this temperature change, however, heat shock gene transcription is activated as shown by the appearance of enhanced levels of mRNAs and proteins encoded by these genes. In addition, within the first few minutes of such treatment, profound changes occur in several cellular compartments and to several cellular macromolecules.

As detected morphologically, the nucleus of the cell becomes much more granular as the chromatin appears to condense and the nucleolar bodies swell. Accompanying these changes are profound effects on DNA and RNA synthesis as detected by uptake of [3H]thymidine and [3H]uridine into stressed cells. For example, DNA synthesis decreases within the first 30 min by greater than 80% and RNA synthesis decreases by greater than 40%. The structure of the DNA nucleosome also is affected as the H2A and H2B histone components are stripped of their ubiquitin moieties (Bond *et al.*, 1988). These activities are indicative of a shift to nonreplicative forms of the DNA and lowered transcriptional activity. With regard to nuclear RNA, there is a block in the processing of the large-molecular-weight precursor to ribosomal RNA. Although synthesis of the latter continues, formation of ribosomes slows because the processed forms of RNA are required for assembly. Other changes that occur in the nucleus of the stressed cell are a block in the splicing of mRNAs and the import into the nucleus of the HSP70, which localizes to the nucleolar bodies.

In the cell cytoplasm, the most obvious morphological changes are the collapse of the intermediate filament network (Fig. 1, top) and conversion of elongated tubelike mitochondria to globular arrays surrounding the nuclear membrane (Fig. 1, bottom). The

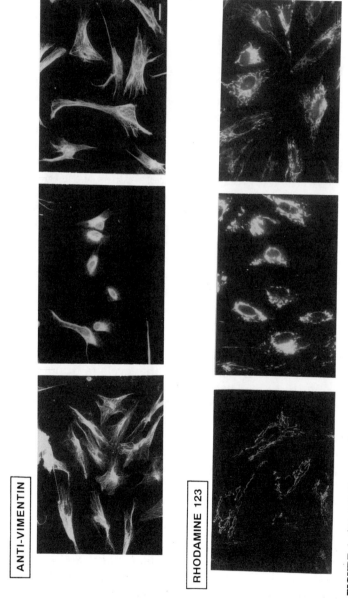

FIGURE 1. (Top) Collapse and recovery of vimentin-containing intermediate filaments after heat shock of chicken embryo fibroblasts. Fibroblasts were fixed with paraformaldehyde and stained with antibodies to vimentin. (Bottom) Collapse and recovery of mitochondria after heat shock of chicken embryo fibroblasts. Mitochondria were visualized by binding of Rhodamine 123 to intact cells prior to fixation. See Collier *et al.* (1993) for experimental details. Reproduced with permission of the publisher.

microtubule and microfilament networks are moderately altered by the stress conditions described and subtle changes occur to portions of the actin-filament system localized to the cell's periphery and in zones corresponding to fibroblast pseudopodia. Actin structures in these regions appear withdrawn and cells shrink in overall shape and volume (Miron et al., 1991).

Metabolic changes that accompany these morphological events include a twofold increase of lactic acid secretion from stressed tissue culture cells, which indicates a shift from respiration to glycolysis for energy production even though levels of the major sources of chemical energy, ATP and phosphocreatinine, are not altered during this degree of stress. The shift to glycolysis may be related to collapse of the mitochondria's morphology for there are also changes in the levels of two of six metabolites of the tricarboxylic acid cycle (the major metabolic activity of mitochondria; Fig. 2). Levels of fumarate and malate decrease by half while pyruvate, citrate, α-ketoglutarate, and succinate remain at the prestressed level. Removal of the stress leads to a relatively rapid return of the normal levels of fumarate and malate and there is concomitantly a restructuring of the mitochondria to their preelongated form. In contrast to these rather mild changes, the levels of aspartate and glutamate rise sharply—the former doubles—during the stress and then return to near normal during recovery. These latter changes may reflect alterations in transmembrane transport systems associated with the mitochondria. In whole animal studies with rats subjected to hyperthermia, the levels of fumarate and malate actually increased in contrast to unaltered levels of citrate, α-ketoglutarate, and succinate. These changes were attributed to increased cerebral oxygen consumption (Goldberg et al., 1966). Other modifications in cellular metabolism include a twofold increase in the amount of proteins conjugated to ubiquitin (the cofactor that targets most intracellular proteins for proteolytic degradation; Fig. 3).

A number of important cellular activities remain unchanged by stress conditions that lead to the alterations noted above. These include functions associated with intracellular trafficking, as posttranslational glycosylations proceed normally on glycoproteins that are transported from the endoplasmic reticulum to the cell's plasma membrane. Also, intracellular levels of inositol trisphosphate and calcium are not altered nor are there detectable changes in intracellular pH.

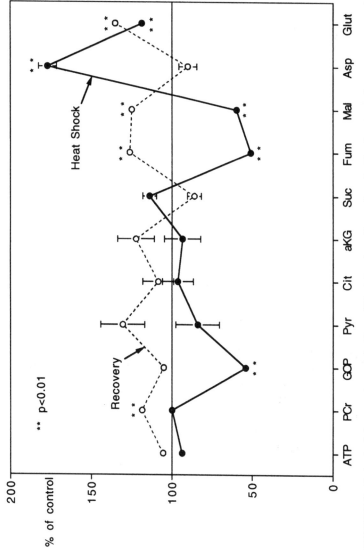

FIGURE 2. Extent of changes in metabolites of the tricarboxylic acid cycle. Triplicate analyses were carried out on duplicate samples of cells treated for 1 hr at 45°C and after a 1-hr recovery at 37°C. Untreated cells were kept at 37°C. See Schlesinger *et al.* (1997) for experimental details.

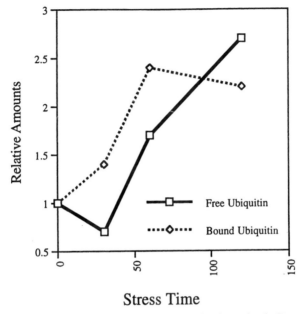

Stress Time

FIGURE 3. Changes in free and bound ubiquitin after heat shock. Stress time is in minutes. See Bond *et al.* (1988) for experimental details.

In the avian cell system, recovery from the stress state is accompanied by the return of normal mRNA transcription, by a reformation of the intermediate filament network, and by the restructuring of the mitochondria; however, new gene transcription appears to be required for all of these (Collier and Schlesinger, 1986b).

The cellular perturbations described above for a physiological hyperthermic stress occur also when cells are treated with 0.1 mM sodium arsenite for 3 hr, both being conditions that do not irreversibly damage the cells.

3. AVIAN STRESS GENES AND ACTIVATION

The major genes activated by thermal stress are closely analogous to those studied in great detail in several other eukaryotic biological systems, most notably *Drosophila*, yeast, and the mouse.

These include an hsp70, an hsp90, an hsp 24, and the polyubi-quitin genes. Another stress gene, the hsp47, encodes a protein that interacts with newly formed collagen polypeptides during their maturation (Nagata et al., 1986). Many additional genes—some encode enzymes of the glycolytic metabolic pathway—respond more modestly to stress. The gene encoding hsp60, a mitochon-drial protein that corresponds to the gro EL (the major heat shock gene of the bacteria), is not extensively activated by heat stress of avian cells.

Activation of these normally quiescent genes by a heat shock occurs when the heat shock factor (a protein termed HSF) binds to a short sequence in the DNA (a contiguous array of the 5-bp module, -GAA- or -TTC-; Perisic et al., 1989). This region is termed the heat shock element and is positioned upstream of the chicken heat shock genes. Heat shock elements and several domains of HSFs are conserved among all eukaryotes; however, most of an HSF sequence is unique to a particular species. The common elements of the HSF include those sequences that interact with the DNA promoter and short domains that form so-called leucine zippers that allow for single polypeptides to form multimeric struc-tures. In the chicken system, there are three isoforms of the HSF; one of these is active only at high temperatures whereas a second one is active in DNA binding only at lower temperatures. The mouse has a similar distribution of HSFs (Sarge et al., 1991). The third isoform in the chicken is restricted to germ cells (Nakai and Morimoto, 1993).

HSFs can exist in a stable, inactive form that is convertible in vitro—by conditions that normally stress a cell—to an active, DNA-binding protein. The inactive HSF is a single polypeptide that has the DNA-binding domain cryptic in contrast to the latter, which exists as a trimer with the DNA-binding domains acces-sible. To explain the stress-induced activation of HSF, it has been proposed that an auxiliary factor maintains the HSF cryptic and that its release allows for oligomerization. The heat shock protein HSP70 has been postulated to be the factor and its removal from the complex with HSF results when nascent polypeptides and other labile proteins in the cell are denatured, thereby providing high-affinity binding sites for HSP70. It is the stress that leads to this change—thus unfolded or misfolded polypeptides would be the sensors of the stressor. In another model, newly synthesized forms of HSF are postulated to form stable, non-DNA-binding monomers under normal conditions but are misfolded during the stress and form complexes with an HSC70 (a constitutively synthesized form of the HSP70). In the presence of ATP, the HSC70 dissociates and DNA-binding HSF trimers form (Schlesinger and Ryan, 1993).

Activation of heat shock genes by stress is a transient response and the shutoff is believed to result from interaction between HSF and HSPs. This negative regulation of hsp genes results from higher levels of HSPs; a strong correlation was found between increasing levels of HSP70 and decreasing levels of HSP70 mRNAs (DiDomenico et al., 1982). The three major heat shock genes of the chicken fibroblast are not induced at the same time. Rather, depending on the stress, there is an ordered response with the hsp90 gene the most sensitive and the hsp24 the least sensitive to stress (Kelley and Schlesinger, 1978). This asynchrony is in sharp contrast with the bacterial heat shock gene response in which all heat shock genes are activated simultaneously.

4. AVIAN HEAT SHOCK PROTEINS

There are four major polypeptides whose synthesis is rapidly induced by stress of avian cells: Their subunit molecular masses are 7, 24, 70, and 90 kDa and the latter three can be found in several isoforms and different oligomeric states. The 24- and 90-kDa proteins are also phosphorylated on serine or threonine residues to varying extents. Isoforms of these polypeptides are found in avian cells growing under normal conditions and the functions they perform under the latter conditions provide important clues for their roles during the stress response. Polyclonal antibodies raised in rabbits immunized with purified preparations of the 70- and 90-kDa chicken proteins cross-react with polypeptides of similar molecular sizes that are induced by stress in a variety of biological species (Table I); thus, the structures of these proteins are strongly conserved throughout the biological world.

The chicken HSP90 is encoded by two distinct genes, alpha and beta, which are the result of gene duplication. The avian beta form is not inducible by thermal stress, which is contrary to the mouse and human HSP90 alpha and beta forms (Meng et al., 1993). The HSP90 protein is an abundant polypeptide in the eukaryotic cell with an ATPase activity. It is autophosphorylated on

TABLE I
Conservation of Heat Shock Protein Structure, Based on Immunoblotting with Antibodies to Chicken Heat Shock Proteins[a]

	Antibody		
Source of cell extract	Anti-p24	Anti-p70	Anti-p89
Yeast			
S. cerevisiae A364A, 24°C	−	+	−
S. cerevisiae A364A, 42°C, 2 hr		++	
Slime mold (*D. discoideum*)	−	+	−
Dinoflagellate (*P. cinctum*)	−	+	−
Corn seedling roots	−	+	−
Worm (*Caenorhabditis elegans*)	−	+	−
Fruit fly			
D. melanogaster embryos, 25°C	−	+	+
D. melanogaster embryos, 30 min, 37°C	−	++	+
Frog			
Xenopus kidney cell line	−	+	+
Xenopus kidney cell line, treated for 24 hr with 50 μM arsenite	−	++	+
Mouse			
L929 cells	−	+	+
L929 cells, treated for 5 hr with 50 μM arsenite		++	+
Human			
WI38 cells	−	+	+
Erythrocyte ghosts	−	−	−

[a]See Kelley and Schlesinger (1982) for experimental details.
Modified with permission from the American Society for Microbiology.

serines and threonines and can bind to HSF, suggesting a role in its autoregulation (see above). It has been found complexed—sometimes only transiently—with a number of other cellular proteins that function in signal transduction pathways. Among these are the sarc tyrosine kinase (Brugge, 1986), serine/threonine kinases such as casein kinase and eIF-2α kinase (Matts *et al.*, 1992), intracellular steroid receptors in their non-DNA-binding state (reviewed in Bohen and Yamamoto, 1994), and two RAS-dependent signal transduction pathways (Doyle and Bishop, 1993). HSP90 also binds to microtubules (Redmond *et al.*, 1989) and the reverse transcriptase of hepatitis B virus (Hu and Seeger, 1996).

HSP70 from chicken fibroblasts occurs in multiple isoforms of which some are made in the absence of stress. Early attempts to purify the protein showed that it was predominantly dimeric in its native state, was mainly cytoplasmic in distribution, but was concentrated in the nucleus of the heat-shocked cell. It is a widely studied heat shock protein and the amino-terminal portion of its structure has been determined to atomic resolution from x-ray diffraction of crystals of the protein (Flaherty *et al.*, 1990). This domain of the protein contains an ATP-binding site and its structure resembles that of other ATP-binding proteins including hexokinase and actin. The carboxyl domain contains a site for binding peptides that are predominantly hydrophobic in sequence.

One of the first clues to its function was the observation that a bovine brain clathrin uncoating ATPase was a constitutively synthesized form of the HSP70 (Chappell *et al.*, 1986). Another normally synthesized form of HSP70 is a protein called BIP, which was identified as a protein that forms a complex with newly synthesized immunoglobulin chains secreted into the lumen of the endoplasmic reticulum (Munro and Pelham, 1986). These data suggested that HSP70 was involved in the folding and unfolding of polypeptides. Detailed genetic analyses in yeast have shown that HSP70 forms a family of 10 proteins: Some function in the transport and folding/unfolding activities essential for importing nuclear-encoded proteins into the inner spaces of the mitochondria. These studies led to the concept that HSP70s are chaperones.

In some systems, HSP70 appears to be complexed with other heat shock proteins to form transportosomes. In general, however, the chaperone concept invokes the model in which proteins susceptible to unfolding and irreversible damage by stress are protected from a degradation pathway through a complexing with the HSP70 or other HSP proteins. In the presence of ATP, this complex is dissociated and the protein allowed to refold in a normal manner. During the stress period, however, the high levels of HSPs would favor continued complex formation.

The chicken HSP24 is an actin-capping protein (Miron *et al.*, 1991), but what its role is during a heat shock response is unclear. A survey of normal adult and embryonic chicken tissue found that all muscle tissue contained 24-kDa proteins that cross-reacted with anti-HSP24 antibodies. In the fibroblast, however, there were very low levels of this protein until the cells were stressed extensively. Heat shock subtly alters actin filaments at the cell's periphery and increased levels of HSP24 would be expected to bind and block actin filament growth as well as protect filaments from disorganization and degradation. Mammalian HSP27 can function as a chaperone to enhance refolding of denatured protein (reviewed in Arrigo and Landry, 1994).

During prolonged stress of the avian fibroblasts, HSP24 molecules accumulate in the cell cytoplasm as granules, which are revealed by electron microscopy to be composed of actinlike fibers (Fig. 4). These aggregates disassemble during the cell's recovery from the stress. When initially isolated from the stressed avian cell, the HSP24 forms a heterogeneous array of oligomeric structures with an octamer as the predominant form and attempts to purify HSP24 led to irreversible protein aggregation. This propensity to form higher-molecular-weight forms accounts for the resemblance of this HSP to the crystallins of the eye. One of the early

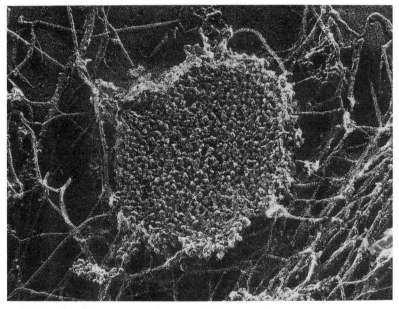

FIGURE 4. Electron micrograph of the HSP24 granule. Sample was prepared by a freeze-fracture, deep-etch, replica formation technique. The diameter of the granule is about 0.5 μm. See Collier *et al.* (1988) for details. Reproduced with permission from Rockefeller University Press.

hints about HSP24 structure was the discovery that the *Drosophila* protein had a short sequence in its genome that was closely homologous to sequences in the mammalian crystallin (Ingolia and Craig, 1982).

The analogous mammalian protein, HSP27, is phosphorylated at two serine sites when cells are activated by mitogens. In an *in vitro* actin-binding system, the phosphorylated and larger-molecular-weight forms of mouse HSP25 do not bind to actin filaments whereas the nonphosphorylated lower oligomeric forms do (Benndorf *et al.*, 1994). These results suggest that HSP24-27 regulates actin filament growth and participates in a signal transduction pathway.

Ubiquitin is a small polypeptide of 78 amino acids that plays a central role in the metabolism of the three major kinds of macromolecules found in all cells—DNA, RNA, and protein (reviewed in Rechsteiner, 1988). All eukaryotic cells contain about 10^6 molecules of ubiquitin and its structure is virtually invariant from plant to human. Most of the polypeptide is transiently covalently bound via its carboxyl-terminal glycine to other proteins in the cell and this interaction affects the function of the target polypeptide. In DNA, ubiquitin alters the structures of the H2A and H2B histones of the nucleosome and its presence occurs primarily during the replicative stage of DNA metabolism—attachment beginning at the S phase of the cell's mitotic cycle and dissociation at completion of DNA replication. In the heat- or chemical-stressed avian fibroblast, there is a rapid release of ubiquitin from H2A and H2B—probably accompanying the condensation of DNA observed in the stressed cell. Ubiquitin's role in RNA metabolism occurs during ribosome formation for one component of the small subunit and one of the large subunit are synthesized with ubiquitin sequences at the amino terminus of the proteins. These fusion proteins are transported into the nucleolus where the ubiquitin portion is removed and the ribosomal proteins become incorporated into the ribosomes. Fused to a ribosomal protein, ubiquitin may act as a stabilizing cofactor; however, its covalent attachment to most cellular proteins initiates their proteolytic degradation. The latter pathway requires a set of ATP-dependent enzymatic reactions plus several unusual proteases organized into a large doughnut-shaped organelle termed the proteasome. The ϵ amino group of exposed lysines in the target protein are ubiquitinated leading to a "tree" of multiple ubiquitin polypeptides. What designates a cellular protein to become a target for ubiquitination is not totally understood, but misfolded and unfolded polypeptides are particularly susceptible.

Hyperthermia and chemical agents increase the level of protein ubiquitination, which indicates that these types of stress are denaturing proteins (Fig. 3). This increase is accompanied by enhanced transcription of genes encoding ubiquitin and enzymes required in the ubiquitin degradation pathway. Ubiquitin itself is encoded in mRNAs that contain multiple copies of ubiquitin, ranging from as few as 3 in some vertebrate species to greater than 50 in invertebrates. The avian ubiquitin genes encode four and five ubiquitins and the latter contains in its 5' regulatory sequences promoters that respond to heat shock transcription factors (Bond and Schlesinger, 1986). Similar polyubiquitin genes controlled by heat shock factors are found in yeast and plants as well. Increased levels of polyubiquitin mRNAs have been measured in tissues obtained from rodents subjected to hyperthermia (Nowak *et al.*, 1990), demonstrating that ubiquitin levels are indicative of stress in both intact animals as well as cells in culture.

A variety of other proteins are altered in stressed avian cells; for example, the total level of the oncoprotein c-myc increases severalfold but much of its regulation is posttranscriptional (Lüscher and Eisenman, 1988). Posttranslational regulation occurs also; among these are changes in the phosphorylation state of several proteins. An overview of the relation of stress proteins to sensitive cellular components is pictured in Fig. 5.

5. TOLERANCE

It has been clearly demonstrated by a variety of experiments in many different kinds of organisms that a brief preexposure to stress leads to an increased ability of the biological system to tolerate additional stress of a more severe quality. This phenomenon of stress tolerance can be correlated in most cases with increased levels of heat shock proteins and the chaperone-type properties of these polypeptides offer a reasonable molecular model to explain tolerance (Parsell and Lindquist, 1994). Importantly, synthesis of heat shock proteins seems to be essential at the elevated temperatures for tolerance to occur, and the inability of an organism to synthesize newly formed chaperonelike proteins

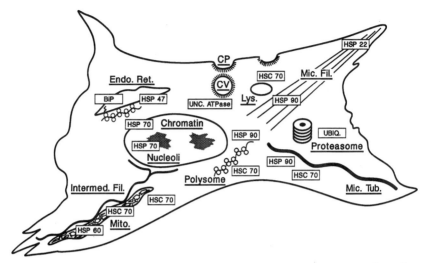

FIGURE 5. A heat shock view of the cell. Reproduced with permission from the International Pediatric Research Foundation, Inc.

seriously affects that organism's survival during a stress. In addition, morphological changes that normally accompany stress such as the collapse of intermediate filaments and mitochondria in avian fibroblasts do not occur in prestressed, tolerant cells and one assumes that these cells are able to continue to function normally over a broad range of physiological conditions. Natural hyperthermic stress in the form of febrile illness or exposure by plants to increasing sunlight tends to occur slowly as opposed to a true "shock." This gradual imposition of stress would allow the cell and organism to initiate the appropriate levels of response— most likely through induction and synthesis of heat shock proteins—so that increased exposure to the stress would not damage the organism.

REFERENCES

Arrigo, A.-P., and Landry, J., 1994, Expression and function of the low-molecular-weight heat shock proteins, in: *The Biology of Heat Shock Proteins and Molecular Chaperones* (R. I. Morimoto, A. Tissieres, and C. Georgopoulos, eds.), Cold Spring Harbor Laboratory Press, Cold Spring Harbor, NY, pp. 335–373.

Atkinson, B. G., 1981, Synthesis of heat-shock proteins by cells undergoing myogenesis, *J. Cell Biol.* **89**:666–671.

Atkinson, B. G., and Dean, R. L., 1985, Effects of stress on the gene expression of amphibian, avian, and mammalian blood cells, in: *Changes in Eukaryotic Gene Expression in Response to Environmental Stress* (B. G. Atkinson and D. B. Whelan, eds.), Academic Press, San Diego, pp. 159–179.

Benndorf, R., Hayess, K., Ryazantsev, S., Wieske, M., Behlke, J., and Lutsch, G., 1994, Phosphorylation and supramolecular organization of murine small heat shock protein HSP25 abolish its actin polymerization-inhibiting activity, *J. Biol. Chem.* **269**:20780–20784.

Bohen, S. P., and Yamamoto, K., 1994, Modulation of steroid receptor signal transduction by heat shock proteins, in: *The Biology of Heat Shock Proteins and Molecular Chaperones* (R. I. Morimoto, A. Tissieres, and C. Georgopoulos, eds.), Cold Spring Harbor Laboratory Press, Cold Spring Harbor, NY, pp. 313–334.

Bond, U., and Schlesinger, M. J., 1986, The chicken ubiquitin gene contains a heat shock promoter and expresses an unstable mRNA in heat-shocked cells, *Mol. Cell. Biol.* **6**:4602–4610.

Bond, U., Agell, N., Haas, A. L., Redman, K., and Schlesinger, M. J., 1988, Ubiquitin in stressed chicken embryo fibroblasts, *J. Biol. Chem.* **263**:2384–2388.

Brugge, J., 1986, Interaction of the Rous sarcoma virus protein pp60src with the cellular proteins pp50 and pp90, *Curr. Top. Microbiol. Immunol.* **123**:1–22.

Chappell, T. G., Welch, W. J., Schlossman, D. M., Palter, K. B., Schlesinger, M.J., and Rothman, J. E., 1986, Uncoating ATPase is a member of the 70 kilodalton family of stress proteins, *Cell* **45**:3–13.

Collier, N. C., and Schlesinger, M. J., 1986a, Induction of heat-shock proteins in the embryonic chicken lens, *Exp. Eye Res.* **42**:103–117.

Collier, N. C., and Schlesinger, M. J., 1986b, The dynamic state of heat shock proteins in chicken embryo fibroblasts, *J. Cell Biol.* **103**:1495–1507.

Collier, N. C., Heuser, J., Levy, M. A., and Schlesinger, M. J., 1988, Ultrastructural and biochemical analysis of the stress granule in chicken embryo fibroblasts, *J. Cell Biol.* **106**:1131–1139.

Collier, N. C., Sheetz, M. P., and Schlesinger, M. J., 1993, Concomitant changes in mitochondria and intermediate filaments during heat shock and recovery of chicken embryo fibroblasts, *J. Cell. Biochem.* **52**:297–307.

DiDomenico, B. J., Bugalsky, G. E., and Lindquist, S., 1982, The heat shock response is self regulated at both the transcriptional and post-transcriptional level, *Cell* **31**:593–603.

Doyle, H., and Bishop, J., 1993, Torso, a receptor tyrosine kinase required for embryonic pattern formation, shares substrates with the sevenless and EGF-R pathways in *Drosophila*, *Genes Dev.* **7**:633–646.

Flaherty, K. M., DeLuca-Flaherty, C., and McKay, D. B., 1990, Three-dimensional structure of the ATPase fragment of a 70K heat-shock cognate protein, *Nature* **346**:623–628.

Goldberg, N. D., Passonneau, J. V., and Lowry, O. H., 1966, Effects of changes in brain metabolism on the levels of citric acid cycle intermediates, *J. Biol. Chem.* **241**:3997–4003.

Hu, J., and Seeger, C., 1996, Hsp90 is required for the activity of a hepatitis B virus reverse transcriptase, *Proc. Natl. Acad. Sci. USA* **93**:1060–1064.

Ingolia, T. D., and Craig, E. A., 1982, Four small *Drosophila* heat shock proteins are related to each other and to mammalian α crystallin, *Proc. Natl. Acad. Sci. USA* **79:**2360–2364.

Kelley, P. M., and Schlesinger, M. J., 1978, The effect of amino acid analogues and heat shock on gene expression in chicken embryo fibroblasts, *Cell* **15:**1277–1286.

Kelley, P. M., and Schlesinger, M. J., 1982, Antibodies to two major heat shock proteins cross react with similar proteins in widely divergent species, *Mol. Cell. Biol.* **2:**267–274.

Lüscher, B., and Eisenman, R. N., 1988, c-myc and c-myb protein degradation: Effect of metabolite inhibitors and heat shock, *Mol. Cell. Biol.* **8:**2504–2512.

Matts, R. L., Xu, A., Pal, J. K., and Chen, J. J., 1992, Interactions of the heme-regulated eIF-2 alpha kinase with heat shock proteins in rabbit reticulocyte lysates, *J. Biol. Chem.*, **267:**18160–18167.

Meng, X., Jerome, V., Devin, J., Baulieu, E. E., and Catelli, M. G., 1993, Cloning of chicken hsp 90 beta: The only vertebrate hsp90 insensitive to heat shock, *Biochem. Biophys. Res. Commun.* **190:**630–636.

Miller, L., and Qureshi, M. A., 1992, Heat-shock protein synthesis in chicken macrophages: Influence of in vivo and in vitro heat shock, lead acetate and lipopolysaccharide, *Poult. Sci.* **71:**988–998.

Miron, T., Vancompernolle, K., Vandekerckhove, J., Wilchek, M., and Geiger, B., 1991, A 25 kD inhibitor of actin polymerization is a low molecular mass heat shock protein, *J. Cell Biol.* **114:**255–261.

Morimoto, R. I., 1993, Cells in stress: Transcriptional activation of heat shock genes, *Science* **259:**1409–1410.

Morimoto, R. I., and Fodor, E., 1984, Cell-specific expression of heat shock proteins in chicken reticulocytes and lymphocytes, *J. Cell Biol.* **99:**1316–1323.

Munro, S., and Pelham, H. R. B., 1986, An hsp70-like protein in the ER: Identity with the 78 kd glucose-regulated protein and immunoglobulin heavy chain binding protein, *Cell* **46:**291–300.

Nagata, K., Saga, S., and Yamada, K. M., 1986, A major collagen-binding protein of chick embryo fibroblasts is a novel heat shock protein, *J. Cell Biol.* **103:**223–229.

Nakai, A., and Morimoto, R. I., 1993, Characterization of a novel chicken heat shock transcription factor, heat shock factor 3, suggests a new regulatory pathway, *Mol. Cell. Biol.* **13:**1983–1997.

Nowak, T. S., Jr., Bond, U., and Schlesinger, M. J., 1990, Heat shock RNA levels in brain and other tissues after hyperthermia and transient ischemia, *J. Neurochem.* **54:**451–458.

Parsell, D. A., and Lindquist, S., 1994, Heat shock proteins and stress tolerance, in: *The Biology of Heat Shock Proteins and Molecular Chaperones* (R. I. Morimoto, A. Tissieres, and C. Georgopoulos, eds.), Cold Spring Harbor Laboratory Press, Cold Spring Harbor, NY, pp. 457–494.

Perisic, O., Xiao, H., and Lis, J., 1989, Stable binding of *Drosophila* heat shock factor to head-to-head and tail-to-tail repeats of a conserved 5 bp recognition unit, *Cell* **59:**797–806.

Rechsteiner, M. (ed.), 1988, *Ubiquitin*, Plenum Press, New York.

Redmond, T., Sanches, E. R., Bresnick, E. H., Schlesinger, M. J., Toft, D. O., Pratt, W. B., and Welsh, M. J., 1989, Immunofluorescence colocalization of the 90-

kDa heat-shock protein and microtubules in interphase and mitotic mammalian cells, *Eur. J. Cell Biol.* **50**:66–75.

Ron, A., and Birkenfeld, A., 1987, Stress proteins in the human endometrium and decidua, *Hum. Reprod.* **2**:277–280.

Sarge, K. D., Zimarino, V., Holm, K., Wu, C., and Morimoto, R. I., 1991, Cloning and characterization of two mouse heat shock factors with distinct inducible and constitutive DNA-binding ability, *Genes Dev.* **5**:1902–1911.

Schlesinger, M. J., and Ryan, C., 1993, An ATP- and hsc70-dependent oligomerization of nascent heat-shock factor (HSF) polypeptide suggests that HSF itself could be a "sensor" for the cellular stress response, *Protein Sci.* **2**:1356–1360.

Schlesinger, M. J., Ashburner, M., and Tissieres, A. (eds.), 1982, *Heat Shock: From Bacteria to Man*, Cold Spring Harbor Laboratory Press, Cold Spring Harbor, NY.

Schlesinger, M. J., Ryan, C., Chi, M.-Y., Carter, J. G., Pusateri, M. E., and Lowry, O. H., 1997, Metabolite changes associated with heat-shocked avian fibroblast mitochondria, *Cell Stress & Chaperones* **2**:25–30.

Voellmy, R., and Bromley, P. A., 1982, Massive heat-shock polypeptide synthesis in late chicken embryos: Convenient system for study of protein synthesis in highly differentiated organisms, *Mol. Cell. Biol.* **2**:479–483.

8

Radiation-Induced Responses in Mammalian Cells

GAYLE E. WOLOSCHAK

In bacteria, a well-characterized response to cellular stress DNA damage has been identified and termed the *SOS response.* This response is a program of gene induction that allows bacteria to respond to damaged DNA and other cellular stresses in a protective manner, allowing for optimal conditions for cell survival (Ossanna *et al.*, 1987). In mammalian cells, a similar DNA-damage response pathway has been shown to exist, but the mechanisms regulated by this pathway are not as clear-cut as in bacteria (Herrlich *et al.*, 1984, 1986; Mai *et al.*, 1989).

In mammalian cells, similar SOS-like responses have been studied in detail. For example, work with induced responses in growth factor and heat shock systems have led to the identification of cellular pathways activated by the cellular agents (Kartasova

GAYLE E. WOLOSCHAK • Center for Mechanistic Biology and Biotechnology, Argonne National Laboratory, Argonne, Illinois 60439.

Stress-Inducible Processes in Higher Eukaryotic Cells, edited by Koval. Plenum Press, New York, 1997.

and van de Putte, 1988). Although similar analyses of radiation-induced responses have lagged somewhat behind, it is clear that the results of these radiation response studies are now contributing to an understanding of the cellular pathways activated following exposure of cells to radiation-induced damage. The purpose of this chapter is to define this radiation-induced response, document the pathways activated by it, and clarify how the activation of these pathways can lead to the consequences of radiation exposure, including cellular transformation and cell death.

1. CELLS IN CULTURE

1.1. Approaches Used

Several different approaches have been used to identify genes induced by radiation and other cellular stresses. Among the earliest was two-dimensional electrophoresis (2DE) of proteins. Total proteins (or proteins in various subcellular locations) were extracted from untreated (control) cells and from cells exposed to stresses such as ultraviolet radiation or ionizing radiation. These proteins were electrophoretically separated on 2DE gels and patterns of protein expression were compared relative to some standard (such as actin). Specific proteins differing in relative accumulation can be identified by comparison to known protein locations, by Western blot, or by microsequencing the protein. This type of approach has been used successfully to identify a large number of radiation-induced genes (Rand and Koval, 1994; Ramsamooj et al., 1992; Boothman et al., 1990; Boothman and Pardee, 1989).

Molecular nonprotein approaches aimed at identifying radiation-induced genes have included differential screening of cDNA libraries, subtractive libraries, and, more recently, differential display of cDNAs. Each of these systems has a variety of advantages and limitations. The choice of approach is based on the needed sensitivity of the assay, the number of test samples to be analyzed in the first screen, and the number of cells available for analysis.

For differential screening of a cDNA library, labeled cDNAs from control and exposed cells are hybridized, and clones showing increased hybridization with the exposed population are selected and characterized. This approach has become more rapid with the use of high-density membranes. It is limited by the sensitivity of the detection system and the fact that transcripts from only two cell types can be compared at the same time. Nevertheless, this approach has contributed much to the identification of radiation-induced genes (Fornace *et al.*, 1988).

Differential display uses PCR for determination of mRNA species differentially expressed in two similar cell populations (Liang and Pardee, 1992, 1993). The procedure uses a $(T)_{12}XY$ oligonucleotide as the 3' primer and an arbitrary 10-mer as the 5' primer. Labeling occurs by inclusion of $\alpha[^{33}P]$-dATP in the PCR. Two artifacts caused by this approach are (1) random priming from dT present from affinity purification of poly(A)$^+$ RNA and (2) hybridization of the arbitrary primer to template target sequences on both cDNA strands. In recent work we have developed a method for avoiding both artifacts: eliminating smearing and identifying nonspecific bands on sequencing gels (Woloschak *et al.*, 1995a). By separately using 5'-end-labeled $(T)_{12}XY$ and arbitrary primers to label bands and comparing two differential display patterns, rather than including labeled nucleotides in the PCR itself, we can distinguish only those products incorporating the $(T)_{12}XY$ primer on the 3' ends and the arbitrary primer on the 5' ends. Those bands that are generated randomly in the PCR are readily detectable and can be ignored. If, on the other hand, one is interested only in a diagnostic banding pattern for differential display, benefit can be derived from the simplicity of the pattern obtained when labeled $(T)_{12}XY$ is used. Each of these approaches has been used successfully for the identification of radiation-induced genes. With the recent availability of cDNA libraries on high-density membranes, differential screening of libraries and differential display coupled with library screening have become efficient and accurate means of identifying full-length copies of radiation-induced genes.

Finally, a combined approach has also been used, though to a limited extent, for the identification of radiation-induced transcripts using *in vitro* translation products as the endpoint to be monitored. In these experiments, RNA from control and radiation-exposed cells is extracted. These RNA preparations are translated *in vitro* using rabbit reticulocyte lysate systems that incorporate labeled amino acids during the translation reaction. The products are analyzed and compared by 2DE. It should be noted that patterns of *in vitro* translation products cannot usually be compared to extracts taken directly from the cells, because the latter are processed (through glycosylation, phosphorylation, and so forth), altering their electrophoretic mobility.

1.2. Induced Responses

1.2.1. Early Response Genes

During the past decade, a large number of studies have identified genes induced in response to DNA-damaging agents such as UV and ionizing radiation (Woloschak and Chang-Liu, 1990, 1991, 1992, 1995; Woloschak *et al.*, 1990a–c, 1994, 1995a–c; Libertin *et al.*, 1994; Sakakeeny *et al.*, 1994; Martin *et al.*, 1993; Anderson and Woloschak, 1992; Herrlich *et al.*, 1992; Fornace, 1992; Ramsamooj *et al.*, 1992; Boothman *et al.*, 1991; Holbrook and Fornace, 1991; Panozzo *et al.*, 1991; Peak *et al.*, 1991; Munson and Woloschak, 1990; Hallahan *et al.*, 1989; Stein *et al.*, 1989a,b; Fornace *et al.*, 1988, 1989a–c; Ronai *et al.*, 1988; Valerie *et al.*, 1988). The collective contribution of these studies has led to the implication that several different transcription/regulatory factors play a key role in the immediate early response, including p53, AP-1, NF-κB, and others (Kharbanda *et al.*, 1995; Sun *et al.*, 1995; Mohan and Meltz, 1994; Nelson and Kastan, 1994; Prasad *et al.*, 1994; Sahijdak *et al.*, 1994; Andalibi *et al.*, 1993; Datta *et al.*, 1992, 1993a,b; Brach *et al.*, 1991; Hallahan *et al.*, 1991a,b; Kastan *et al.*, 1991; McKenna *et al.*, 1991; Angel *et al.*, 1987), and the identification of nuclear and nonnuclear events as playing essential roles in the actual induction process (Koong *et al.*, 1994; Simon *et al.*, 1994; Devary *et al.*, 1993; Hayashi *et al.*, 1993; Uckun *et al.*, 1992; Stein *et al.*, 1989a,b). For some of these transcription factors, target genes in the transcription factor regulon have been identified (Brach *et al.*, 1993; Dominquez *et al.*, 1993; Engstrom *et al.*, 1993; Finco and Baldwin, 1993; Kunsch and Rosen, 1993; Stein *et al.*, 1989b). For example, NF-κB and AP-1 activation contribute to the induction of HIV-LTR following UV exposure (Schreck *et al.*, 1991, 1995; Biswas *et al.*, 1993; Perkins *et al.*, 1993; Kretzchmar *et al.*, 1992; Schmid *et al.*, 1991; Zmudzka and Beer, 1990; Stein *et al.*, 1989a,b; Angel *et al.*, 1987). In addition, AP-1 and NF-κB sites have been found in a large number of UV- and ionizing-radiation-induced genes (Hallahan *et al.*, 1989, 1995a; Sahijdak *et al.*, 1994; Weichselbaum *et al.*, 1994; Vrdoljak *et al.*, 1994; Hiscott *et al.*, 1993;

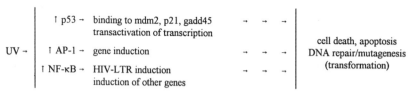

FIGURE 1. UV induces many effects in cells. Projects aimed at identifying proximal events have shown the activation of several pathways following UV exposure.

Messer *et al.*, 1990; Lacoste *et al.*, 1990; Singh and Lavin, 1989). A summary of early gene responses is diagrammed in Fig. 1 and is discussed in more detail below.

1.2.2. Cytokines

A variety of cytokines have been shown to be induced following radiation exposure. A large number of studies have documented the induction of cytokines in response to UV and/or ionizing radiation, including IL-1 (Woloschak *et al.*, 1990b; Schorpp *et al.*, 1984; Ansel *et al.*, 1984), TNF (Chiang and McBride, 1991; Kramer *et al.*, 1990; Weichselbaum *et al.*, 1994), interferons (Woloschak and Liu, 1991), IL-6 (Brach *et al.*, 1993), bFGF (Fuks *et al.*, 1994; Haimovitz-Friedman *et al.*, 1991), β-TGF (Martin *et al.*, 1993; Anscher *et al.*, 1992), vascular endothelial growth factor (Ijichi *et al.*, 1995), and others (Sakakeeny *et al.*, 1994; Datta *et al.*, 1992, 1993a; Peter *et al.*, 1993; Stein *et al.*, 1989d). Some (such as TNF, IL-1) are induced as part of the "early response" during the first 4 hr following exposure, whereas others (like β-TGF) are induced within days following exposure (Anscher *et al.*, 1992). Induction of many of these cytokines are cell type specific. bFGF has been found to be induced following radiation exposure of endothelial cells (Fuks *et al.*, 1994), and Hallahan *et al.* (1995a–c) have documented that pathways of gene induction following ionizing radiation may be different in endothelial cells relative to other cell types. This means that induction following exposure *in vivo* is dependent on studying the proper tissues and cell subpopulations.

The mechanism responsible for cytokine induction appears to be predominantly transcriptional (rather than through release of preformed proteins). As with all radiation-induced genes, nuclear transcription factors (such as NF-κB, AP-1, and p53) are likely to play an important role (see Fig. 1) during the initial portion of the response. The mechanism may be dependent on cell type, cell cycle state, stage of cell transformation, or response state of the cell (discussed below).

The consequences of cytokine induction for the irradiated host are not clear. Neta and co-workers have documented radio-protective effects of some cytokines (Neta et al., 1988, 1989; Neta and Oppenheim, 1988, 1991), many of which work through a bone marrow-sparing effect, whereas others reduce cytotoxicity. The role of cytokines in the cascade of non-radiation-induced cellular responses has been well documented, and it may provide an important functional perspective; cytokines function as effector molecules for pyrogenic responses, immune responses, and cellular defense mechanisms, all of which are important in removing foreign antigens (and perhaps self-antigens generated from radiation exposure) from the host (Chiang and McBride, 1991; Neta and Oppenheim, 1988, 1991; Mossman and Coffman, 1989; Oliff et al., 1987). Cytokines also play an important role in regulating apoptosis. One model of radiation-induced mechanisms suggests that cells that are damaged by radiation but fail to undergo apoptosis become transformed pre-tumor cells; thus, apoptosis of damaged cells is a mechanism to prevent tumorigenesis (Nelson and Kastan, 1994). Cytokines that induce and/or regulate apoptosis, such as IL-1, bFGF, TNF, and others (Chen et al., 1995; Hallahan et al., 1995c; Fuks et al., 1994), may have a protective effect on the host in preventing tumor induction.

1.2.3. Apoptosis-Associated Factors

Many studies have documented that one consequence of radiation exposure is apoptosis (Chen et al., 1995; Fornace, 1992; Uckun et al., 1992; Kastan et al., 1991), and it is therefore likely that the products of many radiation-induced genes will be required for various apoptosis pathways. As noted above, several radiation-induced cytokines (including TNF, IL-1, TGFβ, and IL-6) have been shown to induce apoptosis in vitro, suggesting that one mediator of radiation-induced apoptosis is extracellular (Mangan et al., 1992; Zubiaga et al., 1992; Chiang and McBride, 1991). Recent studies by Fuks et al. (1994) have shown, on the other hand, that bFGF is a negative regulator of apoptosis. It should be noted, however, that both radiation-induced cytokine release and cytokine-specific apoptosis induction are cell type specific and thus cannot adequately explain the broad nature of the induction of apoptosis by radiation. For that reason, many investigators have speculated that radiation directly induces a series of intracellular events that lead to apoptosis.

Several intracellular pathways have been identified as being important in cellular apoptosis, though not all are radiation-induced. For example, Baughman *et al.* (1992) identified a T-cell clone, Tcl 30, that is induced in T cells undergoing glucocorticoid-induced apoptosis but not radiation-induced apoptosis (Woloschak *et al.*, 1996). For radiation-induced apoptosis, p53 has been shown to play a critical role in initiating early events (Nelson and Kastan, 1994; Kastan *et al.*, 1992). Although p53 is not transcriptionally induced following radiation exposure, the protein has been shown to increase quantitatively and functionally in response to radiation (Kastan *et al.*, 1992; Yonish-Rouach *et al.*, 1991). This pathway is essential, for normal cellular response and disruption in some cell types is thought to be critical in the cellular transformation pathway (Sun *et al.*, 1995; Kastan *et al.*, 1992). p53 as an effector of the radiation-induced stress response is discussed in more detail below.

The role of the *myc*/bcl2 pathway in radiation-induced apoptosis is poorly defined. Studies by Chen *et al.* (1995) have shown that radiation causes a decrease in accumulation of Bcl-2, which then leads to radiation-induced apoptosis. Although these apoptosis effectors are an essential component in the stress-induced apoptosis pathway in some cell systems, the role in radiation-induced apoptosis is not clear. Other genes (e.g., *Rp2*, *Rp8*) have been found to be associated with apoptosis and to be radiation-induced, but the pathways that the gene products feed into are unknown (Woloschak *et al.*, 1996; Owens *et al.*, 1991; Sellins and Cohen, 1987).

1.2.4. Growth Arrest Genes

Studies of growth arrest following radiation exposure are numerous, so this section will only review a few key aspects of the literature. Associated with apoptosis is the arrest in cell growth that follows radiation exposure. In general, this delay in cell cycle progression accompanying radiation exposure is associated with the induction of a cascade of gene products that permits the cells to undergo apoptosis or to progress through the cell cycle. Failure to undergo this delay is likely to enhance mutation fixation and tumorigenesis. The G_1 checkpoint is regulated by a p53-dependent pathway; delays at G_2 stage are mediated by c-*myc* (McKenna *et al.*, 1991) and likely by other gene products as well. Precisely how radiation-mediated growth arrest occurs, however, is not clearly understood.

Genes important in regulating the cell cycle have also been shown in numerous studies to be affected by radiation exposure (Muschel *et al.*, 1993; Herrlich *et al.*, 1992; Kastan *et al.*, 1992; McKenna *et al.*, 1991; Boothman *et al.*, 1990; Fornace *et al.*, 1988, 1989a–c). Many of these are genes known to be important in apoptosis or in the early response, as noted above. However, many have as yet unidentified functions, such as many of the gadd genes (growth- arrest-DNA-damage inducible) described by Fornace and co-workers (Kastan *et al.*, 1992; Fornace *et al.*, 1988). Accompanying this arrest in cell cycle progression, downregulation in expression of cell-cycle-related genes is also observed, particularly histones (Sidjanin *et al.*, 1996; Datta *et al.*, 1993a), rRNA (Weber *et al.*, 1990; Fuchs *et al.*, 1990), and others. Boothman *et al.* (1994a) have also characterized radiation-induced transcriptional events associated with cell cycle differences: thymidine kinase, a cell-cycle-regulated gene, was shown to be increased following ionizing radiation exposure of transformed cells.

1.2.5. Viruses

A large number of viruses have been shown to be transcriptionally activated by radiation, including mouse Moloney sarcoma virus (Lin *et al.*, 1990) and SV40 (Vanetti, 1988; Hagedorn *et al.*, 1983). The principle of direct activation of virus expression following cellular stress may account for virus activation following prolonged sunlight exposure, DNA damage, or other environmental stresses. The reasons for such activation are not clear. Moderately repetitive elements such as mouse viruslike 30 S elements, also termed VL30 elements (Panozzo *et al.*, 1991), and long interspersed repetitive elements (Servomaa and Rytomaa, 1990) are both induced by radiation, though the induction of the latter is a later event. This induction may be a remnant of the viral nature of these long repeat elements. VL30 elements have long terminal repeats much like those found in typical retroviruses (Panozzo *et al.*, 1991).

HIV-1 has also been shown in numerous studies to be activated following radiation exposure (Woloschak *et al.*, 1994; Zmudzka and Beer, 1990; Stein *et al.*, 1989a,b; Valerie *et al.*, 1988). A decade ago, Valerie *et al.* (1988) developed a cell line to monitor induction of the HIV promoter through the use of reporter gene expression. HeLa cells stably transfected with an HIV-LTR-CAT[4] construct have been used by several groups to study the ability of DNA-damaging agents to induce expression from the HIV promoter (Valerie and Rosenberg, 1990; Woloschak *et al.*, 1994; Zmudzka and Beer, 1990). Similarly, other *in vitro* and *in vivo* studies have demonstrated induction of HIV-LTR expression after exposure to a variety of DNA-damaging and tumor-promoting agents (Cavard *et al.*, 1990; Morrey *et al.*, 1991; Stanley *et al.*, 1989). The mechanism responsible for this induction involves both NF-κB and AP-1. Induction of other genes in response to DNA-damaging agents has been described in many different systems (Boothman *et al.*, 1991; Fornace, 1992; Fornace *et al.*, 1988, 1989b; Hallahan *et al.*, 1989; Panozzo *et al.*, 1991; Woloschak and Chang-Liu, 1991; Zhan *et al.*, 1994); recent studies have postulated involvement of protein kinase C (Hallahan *et al.*, 1991a; Woloschak *et al.*, 1990a), p53 (Kastan *et al.*, 1991, 1992; Nelson and Kastan, 1994), tyrosine kinases (Uckun *et al.*, 1992, 1993), and other pathways (Stein *et al.*, 1989a; Fornace, 1992) in the modulation of gene expression after radiation exposure.

1.2.6. Cellular Architecture

Recent studies have documented effects of radiation on cytoskeletal elements such as actin and tubulin (Woloschak *et al.*, 1990b,c). Whereas γ-actin mRNA increases in response to radiation, β-actin mRNA accumulation is decreased, effectively altering the ratio of β:γ actin (Woloschak *et al.*, 1990c, 1995c; Woloschak and Chang-Liu, 1991); changes in this ratio have been associated with tumor progression in several different cell systems (Greenberg *et al.*, 1986; Siebert and Fukuda, 1985; Gerstenfeld *et al.*, 1985). Other studies have demonstrated radiation effects on components making up the extracellular matrix such as fibronectin, keratins, and E-selectin (Hallahan *et al.*, 1995b; Woloschak *et al.*,

1995c; Remy *et al.*, 1991) and on enzymes regulating cellular attachment, such as tissue plasminogen activator (Rao *et al.*, 1994; Sawaya *et al.*, 1994; Boothman *et al.*, 1991; Rotem *et al.*, 1987; Ben-Ishai *et al.*, 1984; Miskin and Ben-Ishai, 1981) and collagenase (Angel *et al.*, 1986).

It is not surprising that radiation affects the expression of genes important in maintaining cellular structure, as radiation has been known to cause drastic changes in cell shape (Painter and Young, 1987; Barendsen, 1986) and changes in cellular adhesion properties. It has been suggested that changes in cell adhesion may play a role in regulating mutations or migration of cells *in vivo* (Rao *et al.*, 1993; Boothman *et al.*, 1991). It is not clear, however, whether these shape changes are secondary to the cell cycle effects also induced by radiation exposure.

1.2.7. Gene Induction and Function

Early SOS models for mammalian systems postulated the induction of DNA repair genes in response to radiation exposure. Although cellular defense mechanisms such as the multidrug resistance (*mdr*) gene (Uchiume *et al.*, 1993) and cytokines (described above) are induced following radiation, very few true DNA repair enzymes are part of this response. Of the repair genes induced in response to radiation exposure, β-polymerase is one of the most interesting because of its direct participation in DNA synthesis (Fornace *et al.*, 1989c). Poly(ADP-ribose) polymerase, also induced in response to radiation, is a component of the cellular response to a variety of stresses, including carcinogens (Cleaver and Morgan, 1991; Oleinick and Evans, 1985). The relative contribution of this induction of polyADP-ribose to the cellular response is unclear. Several other genes encoding repair enzymes are also induced following radiation exposure including O6-methylguanine-DNA methyltransferase (Fritz *et al.*, 1991) and RAD18 (Jones and Prakash, 1991), and others (Kaina *et al.*, 1989). The transcript for topoisomerase I has been shown to be repressed following radiation exposure (Boothman *et al.*, 1994b); this decrease may prevent binding of the enzyme to nicked DNA generated during irradiation.

Studies using a priming radiation dose have suggested the existence of a protective mechanism against radiation damage termed the *adaptive response*. Similar responses have been observed in lymphocytes and fibroblasts, and even in whole animals exposed to a low priming radiation dose followed later by a higher challenging dose. This is discussed in more detail in Chapter 9, but it should be noted that studies have implicated specific gene induction as playing a key role in the regulation of the adaptive response (Boothman *et al.*, 1993; Shadley *et al.*, 1987).

How induced gene responses mediate specific radiation-induced functions is not clearly delineated. Several groups have hypothesized that genes induced at low doses will be those important in cell survival and those induced at high doses will be those important in the apoptotic pathway (Woloschak *et al.*, 1991, 1994; Martin *et al.*, 1993). However, many of the identified, DNA-damage-induced genes do not fit into either category or fit into both categories. IL-1, for example, is both a radioprotector and a mediator of apoptosis (Neta and Oppenheim, 1988, 1991; Neta *et al.*, 1989). *c-fos* is important as a signal for cell survival and as a signal for cell death (Vandenberg *et al.*, 1991; Woloschak and Chang-Liu, 1990; Büscher *et al.*, 1988). The importance of other gene induction events such as HLA genes, ornithine decarboxylase, heme oxygenase, and other unidentified genes is not clear (Rosen *et al.*, 1990; Keys and Tyrell, 1989; Lambert *et al.*, 1989; Kartasova *et al.*, 1987). It therefore seems likely that radiation turns on a cascade of events, the balance of which can lead to survival or apoptosis—and it is the balance of this response that determines what the cellular outcome will be. Attempting to categorize the precise role of a gene induction event may be extremely difficult, as pathways in survival and apoptosis are likely to be shared.

1.3. Factors Influencing the Response

1.3.1. Dose, Dose Rates, and Kinetics

Table I lists factors known to influence the radiation-induced gene response. A large body of work has documented the radiation dose dependencies of induced gene responses (e.g., see Fornace, 1992; Herrlich *et al.*, 1992). The induction of most genes in response to radiation forms a typical bell-shaped curve, though the doses for maximal induction of each gene-specific response may vary. For example, whereas γ-actin mRNA in Syrian hamster embryo cells may exhibit a peak induction 1 hr following 100 cGy of γ-rays, the same dose has little effect on TNF gene expression (Woloschak *et al.*, 1990b,c). The peak dose for induction of a par-

TABLE I
Factors Influencing Radiation-Induced Gene Response

Factor	Radiation-induced reponse
Kinetics	Radiation-induced responses are transient
Dose	Bell-shaped curve
Dose rate	Complex response
Radiation type	Different genes induced by UV or ionizing radiations
Radiation quality	Different genes induced by high- or low-LET radiations
Cell cycle stage	Differences in level of response in resting versus cycling cells
Cell type	Differences in genes induced in different cell types
Senescence	Unknown
Cell passage number	Early passage cells induce different genes than late passage
Transformed state	p53 mutant cells do not induce apoptotic pathway
Species	Species-specific differences in induced responses noted

ticular gene may actually be related to the functional significance of the gene product in the radiation/stress response. It has been hypothesized that genes with maximal induction at low doses may be associated with a cellular survival response, whereas genes with maximal induction at high doses may be effectors for apoptosis. Of course, many gene responses show broad dose responses, especially c-fos, c-jun, and γ-actin (Woloschak et al., 1990c, 1995c; Woloschak and Chang-Liu, 1990, 1991; Hallahan et al., 1991a,b; Stein et al., 1989b), suggesting a role for these genes in both the survival and apoptotic pathways.

Although the dose response pattern is apparent in most studies, dose-rate effects have been poorly analyzed in the literature. One difficulty associated with dose-rate experiments is caused by differences in kinetics. Differences in the times required to complete high- and low-dose-rate experiments may contribute significantly to observed differences in gene induction patterns. For a few genes, studies of dose-rate effects have been carried out controlling for all possible effects of kinetics. In these experiments, few differences in expression of cytoskeletal element transcripts could be observed in response to dose-rate differences for both high- and low-LET radiations (Woloschak et al., 1995b,c). Nevertheless, more recent differential display experiments have suggested that a few genes, as yet unidentified, may actually be induced differentially in response to high- and low-dose-rate exposures. Dose-rate effects have also been observed by Komatsu *et al.* (1993) in studies comparing mutation induction and transcription following radiation exposure.

All identified radiation-induced gene changes to date are transient events, occurring at a specific time following exposure and then decreasing some time thereafter. The kinetics (like dose and dose-rate effects) vary with cell type, radiation quality, and in some cases with radiation dose (Woloschak *et al.*, 1990c, 1995c).

1.3.2. Radiation Quality

DNA-damaging agents modulate gene expression, and some of the genes induced or repressed by ionizing radiation are those that have already been shown to be modulated in response to tumor promoters, growth factors, and the like. An interesting corollary concept is the hypothesis that some genes will be similarly modulated by different qualities and types of radiation, whereas other genes will be affected by one quality of radiation and not another. The precise nature of the inducing agent (e.g., oxidative damage, DNA double-strand breaks) is not known. Other work in cellular transformation and DNA repair systems has documented radiation quality effects (Brenner and Hall, 1990; Hill *et al.*, 1987; McWilliams *et al.*, 1983; van der Schans *et al.*, 1983; Han *et al.*, 1980).

Past and current work has identified genes that are repressed after either neutron or γ-ray exposure, including those coding for ornithine decarboxylase and β-actin, and genes induced after neutron or γ-ray exposure, including those for α-tubulin, γ-actin, and IL-1. Table II lists genes induced or repressed following exposure of a single cell type [Syrian hamster embryo (SHE) cells] to different qualities of radiation. In addition, there are many genes induced after γ-ray exposure that are unaffected by or repressed after neutron exposure (c-*fos*, c-*jun*, c-*src*, c-H-*ras*, β-protein kinase C, H4 histone, and VL30 elements) (Woloschak and Chang-Liu, 1990, 1991, 1992, 1995; Woloschak *et al.*, 1990a–c, 1995b,c). Several genes (*Rp8*, α-interferon, and HIV) are induced after neutron, but not after γ-ray, exposure. Most of these experiments have been done in the *in vitro* SHE cell system or in HeLa cells, but some genes (including VL30 elements, c-*src*, c-H-*ras*, c-*fos*, and c-*myc*) have been studied *in vivo* as well (in mice). Recently much of this

TABLE II
Effects of External Radiation Exposure
on Gene Expression in SHE cells[a]

Gene	Radiation quality	
	n_0	γ rays
Interleukin 1	↑	↑
β-Actin	↓	↓
γ-Actin	↑	↑
β-PKC	NC[b]	↑
Rp-8	↑	NC
c-fos	↓	↑
c-myc	NC	NC
α-Tubulin	↑	↑
Fibronectin	↓	↑
Interleukin 6	↑	ND[c]
Proliferating cell nuclear Ag (PCNA)	↑	↑
Superoxide dismutase	ND	↑
c-jun	↑	↑
Rb	↓	↑
H4 histone	↓	↑
p53	NC	NC

[a]All changes evident within the first 4 hr following radiation exposure.
[b]NC, no change.
[c]ND, not done.

work has been done in HeLa cells so that the sequences obtained can be compared with the vast resources of the human genome sequence database.

The reason for the difference in gene modulatory response with radiations of different qualities is not clear. It is possible (as discussed below) that different qualities of radiation induce different types of intracellular damage. Alternatively, high-LET radiation may deposit a locally "high" dose per cell such that the cell never induces a low-dose or "survival" program of gene induction. Experiments are under way to test this model.

Experiments aimed at identifying novel genes induced following UV, neutron, and/or γ-ray exposure have also been initiated. One method is to use a modification of the technique of differential display of cDNAs to allow for increased sensitivity and a lower frequency of artifacts, and such studies have begun to identify genes induced under different conditions of radiation exposure (Liang and Pardee, 1992, 1993; Liang et al., 1992).

1.3.3. Cell Cycle Stage

Radiation inhibits total cellular transcription as well as cell cycle progression (Sidjanin et al., 1996; Woloschak et al., 1990a; Cohn et al., 1984). Progression through the stages of the cell cycle is orchestrated by a battery of cellular regulatory proteins. Expression of these proteins and their regulators is invariably affected by radiation exposure as well. Although it is noted that cells respond to radiation damage with cell cycle arrest, this is actually effected by the induction of signals that interplay with cell cycle modulating products. However, it is also apparent that changes in the stage of the cell cycle also affect radiation-induced gene expression, and therefore, the growth state of the cell culture must be carefully monitored in experiments analyzing radiation-induced gene expression.

Comparisons of the effects of radiation exposure on growth-arrested and cycling SHE cells (Woloschak and Chang-Liu, 1990, 1991) have demonstrated some influence of the cell's proliferative state on radiation-induced gene expression. Cells in different stages of the cell cycle are known to express different gene programs. For example, β-actin and PCNA transcripts are expressed to a higher overall level in cycling cells related to confluent cell cultures (Woloschak and Chang-Liu, 1990, 1991). Radiation exposure affects expression of the genes in different ways. PCNA mRNA levels are not affected by radiation exposure in SHE cells, regardless of the proliferative state of the cell. β-actin mRNA, however, is repressed in resting SHE cells, but slightly induced in cycling cells, following exposure to radiation (Woloschak and Chang-Liu, 1990, 1991). The effects of cell cycle changes must be taken into consideration in all radiation studies (in vitro and in vivo), particularly those monitoring gene expression.

1.3.4. Cell Type

Systematic studies on specific radiation-induced responses in cells of different lineage have not been carried out. The types of experiments are currently limited by the different radiosensitivities of different cell types. Comparisons of radiation-induced gene induction in lymphocytes and in fibroblasts are difficult because doses that induce a predominantly "survival" response in fibroblasts are likely to cause apoptosis in lymphocytes. The gene programs for these induced responses are likely to be dose dependent (as noted above), but many programs show different responses to the same dose in different cell types.

Many of the radiation-induced genes are commonly induced in many different cell types. IL1 has been reported to be induced in fibroblasts, endothelial cells, brain cells, and lymphocytes following radiation exposure (Krämer *et al.*, 1993; Chiang and McBride, 1991; Woloschak *et al.*, 1990b; Ansel *et al.*, 1984; Schorpp *et al.*, 1984). The induction of c-*fos*, β-PKC, and other genes has not been found to be uniform among different cell types; UV radiation, for example, is a potent inducer of c-*fos* expression in fibroblasts (Woloschak *et al.*, 1994) but not in lens cells (Sidjanin *et al.*, 1996). These studies all suggest cell type differences in the radiation-induced response, but uniform controlled experiments are needed to address this issue. *In vivo* work examining multiple tissues from a single animal receiving whole body radiation has addressed this issue (Anderson and Woloschak, 1992; Panozzo *et al.*, 1991) and will be discussed in more detail below.

Species-specific differences in radiation-induced responses are likely to be found, but again have not been well characterized. The fact that repair pathways are different among some different species supports the possibility of different radiation responses (Applegate and Ley, 1987; Wade and Trosko, 1983). It is generally accepted that species differences in the radiation-induced gene response exist, but experiments comparing the same cell type from different mammalian species with similar repair capacities have not been done.

1.3.5. Age and Culture of Cells

Few studies have examined effects of radiation on cells derived from young versus old animals or on cells from primary cultures versus senescent cultures. Many theories of radiation effects have noted similarities between cells exposed to radiation and senescent cells in culture (Dice, 1993; Cutler and Semsei, 1989; Warner and Price, 1989). It has been proposed that radiation exposure serves as an agent to induce or promote aging, and many features characteristic of radiation-exposed cells (reduced transcription rate, mutation induction, reduced metabolic activities) are also observed in senescent cell cultures.

The effects of the number of cell passages on radiation-induced gene expression have been noted, but these studies have focused only on effects in hamster cells (Woloschak *et al.*, 1990a–c, 1994; Chang-Liu and Woloschak, unpublished information). This work has demonstrated little effect of radiation on gene expression at early passages (passages 1–24) for both high and low doses. However, at later passages, several differences are noted: (1) the cells at later passages, in the presence or absence of radiation, display a generally higher level of expression of cell-cycle-related genes such as PCNA; (2) the levels of most genes that are modulated in response to radiation are affected similarly in early- and later-passage cultures of cells; but (3) p53 mRNA was found to be radiation-induced in later-passage cells but not in cells from earlier passages. This latter finding may be related to oncogenic changes associated with cell passage, though changes in p53 expression following radiation exposure have been reported in untransformed cells (Maltzman and Czyzyk, 1984). These results clearly suggest that cell passage can influence radiation-induced gene expression. How this relates to senescence is not apparent; further studies are needed in this area.

The stage of transformation of the cell has also been shown to influence gene induction indirectly. SHE cells with mutant p53 induce at least three new genes in response to radiation relative to SHE cells with normal p53 function (Chang-Liu and Woloschak, unpublished data). The fact that p53 mutations alter the radiation-induced apoptotic response of cells (Kessis *et al.*, 1993; Kastan *et al.*, 1991) further supports this observed difference. Nevertheless, a systematic study of how mutations/alterations in specific oncogenes affect the radiation-induced response has not been carried out.

1.4. Mechanisms of Response Induction

1.4.1. Inducing Signal

What is the intracellular signal responsible for radiation-induced gene expression? The fact that different qualities and types of radiation induce not entirely overlapping subsets of genes suggests that the intracellular signal may be complex. For example, differences in gene induction patterns observed following exposure to UVA and UVB radiations may be attributed to differences in the types of cellular damage elicited: UVA radiation predominantly causes cellular membrane damage, whereas UVB radiation damages DNA. These two qualities of radiation induce some genes in common but others differently (Sidjanin *et al.*, 1996; Peak *et al.*, 1991), suggesting that some genes are induced by DNA damage. This hypothesis was verified in studies by Devary *et al.* (1993) demonstrating that nuclear signals were not required for activation of src kinase following UV irradiation.

Although both DNA damage and membrane damage induce different programs of genes, it is also possible that the cell distinguishes different types of DNA damage for the inducing signal. As noted above, cells exposed to high- or low-LET radiations induce many common but also some different sets of genes (Table II); both cause predominantly membrane damage. The differences in these responses are not clearly understood, but it is possible that (1) different types of DNA damage are elicited by the different exposures, (2) cellular compartments are affected differently by high- and low-LET radiations, and (3) protein/lipid damage may differentially play a role in the response. Although it is clear that DNA and membrane damage can each serve as intracellular inducers of gene expression, it is also likely that other types of cellular damage play a role.

Recent studies have documented that many radiation-induced transcription effects can be mimicked by conditions under which reactive oxygen intermediates are generated. This has been particularly true for the NF-κB pathway, which appears to be regulated through oxidoreduction (Koong *et al.*, 1994; Hayashi *et al.*, 1993; Schreck *et al.*, 1991) and is also induced with radiation exposure. Prasad *et al.* (1994) have documented that particularly low doses of ionizing radiation are capable of activating NF-κB, and others have demonstrated that exposure to UV radiation also activates an NF-κB response (Woloschak *et al.*, 1995d; Simon *et al.*, 1994; Devary *et al.*, 1993). This suggests that reactive oxygen intermediates induced intracellularly following radiation exposure may induce gene expression through NF-κB activation.

1.4.2. Signal Transduction

The molecular sensor in the cell for radiation effects is some sort of cellular damage, which induces a cascade of events in the damaged cell. This cascade is dependent on factors defined above (Table I), such as cell cycle stage and quality of radiation. Following radiation exposure, a series of phosphorylation events occurs that leads to the activation of transcription of radiation-induced genes (Devary *et al.*, 1993; Uckun *et al.*, 1992). This phosphorylation cascade appears to be complex and involves stress-activated protein kinase (SAPK), as well as mitogen-activated protein kinase (MAPK), pathways. It has been noted that UV responses and x-ray stresses are generally strong inducers of SAPK, and weak inducers of MAPK, pathways (Kharbanda *et al.*, 1995; Uckun *et al.*, 1992). However, other strong inducers of the SAPK pathway include hypo-osmolarity, ceramide, heat shock, protein synthesis inhibition, peroxide, TNF, and IL-1β; the fact that these gene inducers are not identical to x rays or UV radiation when comparing radiation-induced gene sets suggests that subsets of the SAPK pathway or other pathways are also involved in the radiation-induced response.

Protein kinase C (PKC) pathways have also been implicated in radiation-induced gene effects. Woloschak *et al.* (1990a) demonstrated transcriptional induction of β-PKC following exposure of SHE cells to γ rays. Matsui and DeLeo (1990) and Peak *et al.* (1991) demonstrated induction of PKC in human cells in response to sunlight exposure. Hallahan *et al.* (1991a,b) established that PKC mediates the x-ray inducibility of *egr1* and *jun*, and later studies documented PKC mediation in the induction of TNF following ionizing radiation exposure (Hallahan *et al.*, 1993). Recent studies by Hallahan's group, using inhibitors of phospholipase A2, have demonstrated that radiation-induced TNF expression requires phospholipase A2 for induction (Hallahan *et al.*, 1994a,b). Moreover, intracellular calcium is a component of this pathway (Hallahan *et al.*, 1994b,c). Other studies, however, showed that PKC was not required for the induction of the gadd45 gene response to ionizing radiation (Papathanasiou *et al.*, 1991). A relationship has been suggested between PKC activation and NF-κB (Dominquez *et al.*, 1993), which may also be mediated through reactive oxygen intermediates generated in response to radiation exposure.

Other kinases with unknown specificities have been identified as being induced following radiation exposure (Sakuma *et al.*, 1995; Ben-Ishai *et al.*, 1990). Knowledge of the relative contributions of these inductive events will be essential in determining the signal pathway regulated in response to radiation exposure.

1.4.3. Transcriptional Changes

One outcome of the activation of the various kinase cascades is the activation of transcription factors and the induction of new gene transcription. Although changes in mRNA accumulation can be effected by altered mRNA stability, altered rates of transport of transcripts from the nucleus to the cytoplasm, or altered transcription rates, most changes in mRNA accumulation caused by radiation exposure are associated with changes in transcription rates. A few exceptions to this, however, have been noted in the literature (Hilgers *et al.*, 1991). Regulation of specific transcription factor binding, then, is a major early event that leads to the induction of new gene expression in radiation exposed cells (Krämer *et al.*, 1990). This activation leads to other downstream events. Regulation of transcription is mediated by transcription factor binding, but features such as chromatin state, methylation state, and others play a role (Lieberman *et al.*, 1983).

NF-κB is one of the major transcription factors activated in response to UV and ionizing radiation, as noted above (Woloschak *et al.*, 1995d; Brach *et al.*, 1991, 1993). This activation is mediated predominantly by radiation-induced membrane damage and not DNA damage (Simon *et al.*, 1994; Devary *et al.*, 1993). The activation of NF-κB is regulated by its inhibitor IκB, which can bind to NF-κB and prevent its binding to the target DNA nucleotide sequence (Liou and Baltimore, 1993). IκB is regulated by phosphorylation, which can be inhibited *in vitro* with such drugs as pentoxifylline (Biswas *et al.*, 1993) and salicylic acid (Woloschak *et al.*, 1995d; Frantz and O'Neill, 1995; Kopp and Ghosh, 1994). IκB phosphorylation is regulated, at least in part, by the c-*raf* gene product (Finco and Baldwin, 1993; Kasid *et al.*, 1993), which has been associated with radioresistance. Hallahan *et al.* (1995b) have shown that, in endothelial cells, NF-κB binding is also induced in response to radiation exposure. NF-κB binding is necessary (but not sufficient) for radiation-induced transcription, though the other factors required have not been clearly identified. Perkins *et al.* (1993) have shown an association between Sp1 and NF-κB binding that is required for HIV transcriptional induction. Other factors are likely to be important as well.

The AP-1 transcription factor is made up of homo- and heterodimers of c-*fos* and c-*jun* gene products, which are self-regulatory and radiation-induced (Rozek and Pfeifen, 1993; Stein *et al.*,

1989b, 1992; Hollander and Fornace, 1989; Büscher *et al.*, 1988). In contrast to NF-κB, which responds to membrane damage, one intracellular signal for AP-1 binding to its recognition sequence is DNA damage (Stein *et al.*, 1989b). The pathways for phorbol ester-mediated and UV-mediated gene induction converge at AP-1 binding to its recognition sequence. This explains why the UV and phorbol ester responses share many overlapping genes in their regulons. Other genes included in the AP-1 regulon are listed in Table III. It should be noted that HIV-1 is regulated by the AP-1 and NF-κB regulons; the nature of this dual regulation is currently the focus of intense investigation. Interactions between p53 (required for DNA damage recognition) and AP-1 transcription factor binding are also not yet apparent.

Other transcription factors/DNA-binding proteins have been observed to play a role in radiation-induced responses but have been less intensively studied. Ronai's laboratory has identified a UV-response element (URE) and binding protein associated with induction of polyoma virus expression and viral replication (Ronai *et al.*, 1990, 1994; Perkins *et al.*, 1993; Yang *et al.*, 1993; Rutberg *et al.*, 1992; Ronai and Weinstein, 1990). Singh and Lavin (1989) also identified a DNA-binding protein with a specific recognition sequence that is actively induced by γ-ray exposure; the genes regulated in response to this binding have not been identified. Boothman's laboratory has identified increased binding of Sp1, NF-κB, and CREB transcription factors following radiation exposure,

TABLE III
Radiation-Induced Regulons

Transcription factor/regulon	Sequence	Radiation-induced target genes
NF-κB	GGGACTTTCC	IL1, TNF-∞, TNF-β, interferons, IL6, IL8, HIV LTR, major histocompatibility antigens class I, TGF-β, E-selectins
AP-1	ATGAGTCAGCC	Metallothionein, collagenase, c-*fos*, c-*jun*, HIV
URE	TGACAACA	Polyoma virus
CrG	CC(A/T)GGG	*egr*-1, HIV-1
γ ray inducible	TGTCAGTTAGGGT	Unknown

though specific genes have not been fully characterized (Booth-man *et al.*, 1994a–c; Sahijdak *et al.*, 1994).

2. ANIMAL STUDIES

Whereas many studies have documented effects of radiation exposure, such as radiation-induced tumorigenesis, in whole animal systems, few have analyzed radiation effects on gene transcription *in vivo*. The studies that have been conducted were essentially used to establish that identified *in vitro* responses occur *in vivo* as well.

Most whole animal studies have been carried out using inbred mouse strains. Munson and Woloschak (1990) demonstrated that the transient transcriptional arrest that occurs within the first hour after radiation exposure of cultured cells is also evident in mice receiving whole body irradiation. When specific genes were analyzed for transcriptional induction, tissue-specific (Anderson and Woloschak, 1992) and radiation quality differences (Panozzo *et al.*, 1991) were evident, as had been found *in vitro*. Studies have documented that whole body radiation exposure induces TGF-β (Anscher *et al.*, 1992), c-*fos* (Munson and Woloschak, 1990), c-H-*ras* (Anderson and Woloschak, 1992), and VL30 (Panozzo *et al.*, 1991). These events are early, occurring within the first 5 min to 24 hr following radiation exposure. More long-term transcriptional changes associated with fibrosis, particularly increased accumulation of collagen (McAnulty *et al.*, 1991), have been observed. Working with pigs, Martin *et al.* (1993) described radiation-mediated transcriptional induction of TGF-β and β-actin. Later changes in expression of collagen were also observed (Remy *et al.*, 1991); these changes were associated with later induction of fibrosis.

The increasing use of transgenic systems has permitted evaluation of the relative contribution of specific genes and pathways to endpoints such as tumorigenesis, animal death, and others. The application of these transgenic models to studies of radiation-induced gene expression will provide important insights into the relative contribution of gene regulatory events to specific function.

ACKNOWLEDGMENTS. This work was supported by the U.S. Department of Energy, Office of Health and Environmental Research, under Contract No. W-31-109-ENG-38 and NIH grant # ES07141-02. The author thanks Tatjana Paunesku, Frank Collart, and David Grdina for critical review of the manuscript and Ms. Felicia King for outstanding secretarial assistance.

REFERENCES

Andalibi, A., Liao, F., Imes, S., Fogelman, A. M., and Lusis, A. J., 1993, Oxidized lipoproteins influence gene expression by causing oxidative stress and activating the transcription factor NF-κB, *Biochem. Soc. Trans.* **21:**651–655.

Anderson, A., and Woloschak, G. E., 1992, Cellular proto-oncogene expression following exposure of mice to γ-rays, *Radiat. Res.* **130:**340–344.

Angel, P., Poting, A., Mallick, U., Rahmsdorf, H. J., Schorpp, M., and Herrlich, P., 1986, Induction of metallothionein and other mRNA species by carcinogens and tumor promoters in primary human skin fibroblasts, *Mol. Cell. Biol.* **6:** 1760–1766.

Angel, P., Imagawa, M., Chiu, R., Stein, B., Imbra, R. J., Rahmsdorf, H. J., Jonat, C., Herrlich, P., and Karin, M., 1987, Phorbol ester-inducible genes contain a common *cis* element recognized by a TPA-modulated *trans*-acting factor, *Cell* **49:**729–739.

Anscher, M. S., Crocker, J. R., and Jirtle, R. L., 1992, Transforming growth factor-β-1 expression in irradiated liver, *Radiat. Res.* **122:**77–85.

Ansel, J., Luger, T. A., Kock, A., Hochstein, D., and Green, I., 1984, The effect of in vitro UV irradiation on the production of Il1 by murine macrophages and P388D₁ cells, *J. Immunol.* **133:**1350–1355.

Applegate, L. A., and Ley, R. D., 1987, Excision repair of pyrimidine dimers in marsupial cells, *Photochem. Photobiol.* **45:**241–245.

Barendsen, G. W., 1986, Effects of radiation on the reproductive capacity and proliferation of cells in relation to carcinogenesis, in: *Radiation Carcinogenesis* (R. E. Upton, F. J. Albert, and R. E. Shore, eds.), Elsevier, Amsterdam, pp. 85–106.

Baughman, G., Lesley, J., Trotter, J., Hyman, R., and Bourgeois, S., 1992, Tcl-30, a new T cell-specific gene expressed in immature glucocorticoid-sensitive thymocytes, *J. Immunol.* **149:**1488–1496.

Ben-Ishai, R., Sharon, R., Rothman, M., and Miskin, R., 1984, DNA repair and induction of plasminogen activator in human fetal cells treated with ultraviolet light, *Carcinogenesis* **5:**357–362.

Ben-Ishai, R., Scharf, R., Sharon, R., and Kapten, I., 1990, A human cellular sequence implicated in trk oncogene activation is DNA damage inducible, *Proc. Natl. Acad. Sci. USA* **87:**6039–6043.

Biswas, D. K., Dezube, B. J., Ahlers, C. M., and Pardee, A. B., 1993, Pentoxifylline inhibits HIV-1 LTR-driven gene expression by blocking NFκB action, *J. AIDS* **6:**778–786.

Boothman, D. A., and Pardee, A. B., 1989, Inhibition of radiation-induced neoplastic transformation by beta lapachone, *Proc. Natl. Acad. Sci. USA* **86:**4963–4967.

Boothman, D. A., Lee, S., Trask, D. K., Dou, Q-P., and Hughes, E. N., 1990, X-ray-inducible proteins and genes in human cells, in: *Ionizing Radiation Damage to*

DNA: Molecular Aspects (S. S. Wallace and R. B. Painter, eds.), Wiley–Liss, New York, pp. 309–317.

Boothman, D. A., Wang, M., and Lee, S. W., 1991, Induction of tissue-type plasminogen activator by ionizing radiation in human malignant melanoma cells, *Cancer Res.* **51**:5587–5595.

Boothman, D. A., Meyers, M., Fukunaga, N., and Lee, S. W., 1993, Isolation of x-ray-inducible transcripts from radioresistant human melanoma cells, *Proc. Natl. Acad. Sci. USA* **90**:7200–7204.

Boothman, D. A., Davis, T. W., and Sahijdak, W. M., 1994a, Enhanced expression of thymidine kinase in human cells following ionizing radiation, *Int. J. Radiat. Biol.* **30**:391–398.

Boothman, D. A., Fukunaga, N., and Wang, M., 1994b, Down-regulation of topoisomerase I in mammalian cells following ionizing radiation, *Cancer Res.* **54**: 4618–4626.

Boothman, D. A., Lee, I. W., and Sahijdak, W. M., 1994c, Isolation of an x-ray-responsive element in the promoter region of tissue-type plasminogen activator: Potential uses of x-ray-responsive elements for gene therapy, *Radiat. Res.* **138**:S68–S71.

Brach, M. A., Hass, R., Sherman, M. L., Gunji, H., Weichselbaum, R., and Kufe, D., 1991, Ionizing radiation induces expression and binding activity of the nuclear factor κB, *J. Clin. Invest.* **88**:691–695.

Brach, M. A., Gruss, H. J., Kaisho, T., and Asano, Y., 1993, Ionizing radiation induces expression of interleukin 6 by human fibroblasts involving activation of nuclear factor kappa B, *J. Biol. Chem.* **268**:8466–8472.

Brenner, D. J., and Hall, E. J., 1990, The inverse dose-rate effect for oncogenic transformation by neutrons and charged particles: A plausible interpretation consistent with published data, *Int. J. Radiat. Biol.* **58**:745–758.

Büscher, M., Rahmsdorf, H. J., Litfin, M., Karin, M., and Herrlich, P., 1988, Activation of the c-*fos* gene by UV and phorbol ester: Different signal transduction pathways converge to the same enhancer element, *Oncogene* **3**:301–311.

Cavard, C., Zider, A., Vernet, M., Bennon, M., Saragosti, S., Grimber, G., and Briand, P., 1990, In vivo activation by ultraviolet rays of the human immunodeficiency virus type 1 long terminal repeat, *J. Clin. Invest.* **86**:1369–1374.

Chen, M., Quintans, J., Fuks, Z., Thompson, C., Kufe, D. W., and Weichselbaum, R. R., 1995, Suppression of Bcl-2 messenger RNA production may mediate apoptosis after ionizing radiation, tumor necrosis factor α, and ceramide, *Cancer Res.* **55**:991–994.

Chiang, C. S., and McBride, W. H., 1991, Radiation enhances tumor necrosis factor α production by murine brain cells, *Brain Res.* **566**:265–269.

Cleaver, J. E., and Morgan, W. F., 1991, Poly(ADP-ribose)polymerase: A perplexing participant in cellular responses to DNA breakage, *Mutat. Res.* **257**:1–18.

Cohn, S. M., Krawisz, B. R., Dresler, S. L., and Lieberman, M. W., 1984, Induction of replicative DNA synthesis in quiescent human fibroblasts by DNA damaging agents, *Proc. Natl. Acad. Sci. USA* **81**:4828–4832.

Cutler, R. G., and Semsei, I., 1989, Development, cancer and aging: Possible common mechanisms of action and regulation, *J. Gerontol.* **44**:25–34.

Datta, R., Rubin, E., Sukhatme, V., Qureshi, S., Hallahan, D., Weichselbaum, R. R., and Kufe, D. W., 1992, Ionizing radiation activates transcription of the EGR1 gene via CArG elements, *Proc. Natl. Acad. Sci. USA* **89**:10149–10153.

Datta, R., Taneja, N., Sukhatme, V., Qureshi, S. A., Weichselbaum, R., and Kufe,

D. W., 1993a, Reactive oxygen intermediates target CC(A/T)GGG sequences to mediate activation of the early growth response 1 transcription factor gene by ionizing radiation, *Proc. Natl. Acad. Sci. USA* **90**:2419–2422.

Datta, R., Weichselbaum, R., and Kufe, D. W., 1993b, Ionizing radiation down-regulates histone H1 gene expression by transcriptional and post-transcriptional mechanisms, *Radiat. Res.* **133**:176–181.

Devary, Y., Rosette, C., DiDonato, J. A., and Karin, M., 1993, NFκB activation by ultraviolet light not dependent on a nuclear signal, *Science* **261**:1442–1445.

Dice, J. F., 1993, Cellular and molecular mechanisms of aging, *Physiol. Rev.* **73**: 149–159.

Dominquez, I., Sanz, L., Arenzana-Seisdedos, F., and Diaz-Meco, M. T., 1993, Inhibition of protein kinase C zeta subspecies blocks the activation of an NF-κB-like activity in *Xenopus* laevis oocytes, *Mol. Cell. Biol.* **13**:1290–1295.

Engstrom, Y., Kadalayil, L., Sun, S.-C., Samakovlis, C., Hultmark, D., and Faye, I., 1993, κB-like motifs regulate the induction of immune genes in *Drosophila*, *J. Mol. Biol.* **232**:327–333.

Finco, T. S., and Baldwin, A. S., 1993, κB site-dependent induction of gene expression by diverse inducers of nuclear factor κB requires Raf-1, *J. Biol. Chem.* **268**:17676–17679.

Fornace, A. J., Jr., 1992, Mammalian genes induced by radiation: activation of genes associated with growth control, *Ann. Rev. Genet.* **26**:507–526.

Fornace, A. J., Jr., Alamo, I., Jr., and Hollander, C. M., 1988, DNA damage-inducible transcripts in mammalian cells, *Proc. Natl. Acad. Sci. USA* **85**:8800–8804.

Fornace, A. J., Jr., Alamo, I., Hollander, C. M., and Lamoreaux, E., 1989a, Ubiquitin RNA is a major stress-induced transcript in mammalian cells, *Nucleic Acids Res.* **17**:1215–1230.

Fornace, A. J., Jr., Fargnoli, J., Papathanasiou, M., Holbrook, N. J., Hollander, C. M., Nebert, D. W., and Luethy, J. D., 1989b, Mammalian genes coordinately regulated by growth arrest signals and DNA-damaging agents, *Mol. Cell. Biol.* **9**:4196–4203.

Fornace, A. J., Jr., Zmudzka, B., Hollander, C. M., and Wilson, S. H., 1989c, Induction of β-polymerase mRNA by DNA-damaging agents in Chinese hamster ovary cells, *Mol. Cell. Biol.* **9**:851–853.

Frantz, B., and O'Neill, E. A., 1995, The effect of sodium salicylate and aspirin on NF-κB, *Science* **270**:2017–2019.

Fritz, G., Tano, K., Mitra, S., and Kaina, B., 1991, Inducibility of the DNA repair gene encoding O^6-methylguanine-DNA methyltransferase in mammalian cells by DNA damaging treatments, *Mol. Cell. Biol.* **11**:4660–4668.

Fuchs, P., Krolak, J. M., McClain, D., and Minton, K. W., 1990, 18S RNA degradation is not accompanied by altered rRNA transport at early times following irradiation of HeLa cells, *Radiat. Res.* **121**:67–70.

Fuks, Z., Persaud, R. S., Aifiere, A., McLoughlin, M., Ehleiter, D., Schwartz, J. L., Seddon, A. P., Cordon-Cardo, C., and Haimovitz-Friedman, A., 1994, Basic fibroblast growth factor protects endothelial cells against radiation-induced programmed cell death *in vitro* and *in vivo*, *Cancer Res.* **54**:2582–2590.

Gerstenfeld, L. C., Finer, M. H., and Boedtker, H., 1985, Altered β-actin gene expression in phorbol myristate acetate-treated chondrocytes and fibroblasts, *Mol. Cell. Biol.* **5**:1425–1433.

Greenberg, M. E., Hermanowski, A. L., and Ziff, E. B., 1986, Effect of protein synthesis inhibitors on growth factor activation of c-*fos*, c-*myc*, and actin gene transcription, *Mol. Cell. Biol.* **6**:1050–1057.

Hagedorn, R., Thielmann, H. W., Fischer, H., and Schroedes, C. H., 1983, SV40-induced transformation and T-antigen production is enhanced in normal and repair-deficient human fibroblasts after pretreatment of cells with UV light, *J. Cancer Res. Clin. Oncol.* **106**:93–96.

Haimovitz-Friedman, A., Vlodavsky, I., Chaudhuri, A., Witte, L., and Fuks, Z., 1991, Autocrine effects of fibroblast growth factor in repair of radiation damage in endothelial cells, *Cancer Res.* **51**:2552–2558.

Hallahan, D. E., Spriggs, D. R., Beckett, M. A., Kufe, D. W., and Weichselbaum, R. R., 1989, Increased tumor necrosis factor α mRNA after cellular exposure to ionizing radiation, *Proc. Natl. Acad. Sci. USA* **86**:10104–10107.

Hallahan, D. E., Sukhatme, V. P., Sherman, M. L., Virudachalam, S., Kufe, D. W., and Weichselbaum, R. R., 1991a, Protein kinase C mediates X-ray inducibility of nuclear signal transducers *egr-1* and Jun, *Proc. Natl. Acad. Sci. USA* **88**: 2156–2160.

Hallahan, D. E., Virudachalam, S., Sherman, M. L., Huberman, E., Kufe, D. W., and Weichselbaum, R. R., 1991b, Tumor necrosis factor gene expression is mediated by protein kinase C following activation by ionizing radiation, *Cancer Res.* **17**:4565–4569.

Hallahan, D. E., Gius, D., Kuchibhotla, J., Sukhatme, V., Kufe, D. W., and Weichselbaum, R. R., 1993, Radiation signaling mediated by Jun activation following dissociation from a cell type-specific repressor, *J. Biol. Chem.* **268**: 4903–4907.

Hallahan, D. E., Virudachalam, S., Kufe, D. W., and Weichselbaum, R. R., 1994a, Ketoconazole attenuates radiation-induction of tumor necrosis factor, *Int. J. Radiat. Oncol. Biol. Phys.* **29**:777–780.

Hallahan, D. E., Virudachalam, S., Kuchibhotla, J., Kufe, D. W., and Weichselbaum, R. R., 1994b, Membrane-derived second messenger regulates x-ray-mediated tumor necrosis factor alpha gene induction, *Proc. Natl. Acad. Sci. USA* **91**: 4897–4901.

Hallahan, D. E., Bleakman, D., Virudachalam, S., Lee, D., Grdina, D., Kufe, D. W., and Weichselbaum, R. R., 1994c, The role of intracellular calcium in the cellular response to ionizing radiation, *Radiat. Res.* **138**:392–400.

Hallahan, D. E., Dunphy, E., Virudachalam, S., Sukhatme, V. P., Kufe, D. W., and Weichselbaum, R. R., 1995a, c-*jun* and *Egr-1* participate in DNA synthesis and cell survival in response to ionizing radiation exposure, *J. Biol. Chem.* **270**: 30303–30309.

Hallahan, D. E., Clark, E. T., Kuchibhotla, J., Gewertz, B. L., and Collins, T., 1995b, E-selectin gene induction by ionizing radiation is independent of cytokine induction, *Biochem. Biophys. Res. Commun.* **217**:784–795.

Hallahan, D. E., Kufe, D. W., and Weichselbaum, R. R., 1995c, Spatial and temporal control of gene therapy using ionizing radiation, *Nature Med.* **1**:786–791.

Han, A., Hill, C. K., and Elkind, M. M., 1980, Repair of cell killing and neoplastic transformation at reduced dose rates of ^{60}Co γ rays, *Cancer Res.* **40**:3328–3332.

Hayashi, T., Ueno, Y., and Okamoto, T., 1993, Oxidoreductive regulation of NF-κB. Involvement of a cellular reducing catalyst thioredoxin, *J. Biol. Chem.* **268**: 11380–11388.

Herrlich, P., Mallick, U., Ponta, H., and Rahmsdorf, H. J., 1984, Genetic changes in mammalian cells reminiscent of an SOS response, Hum. Genet. **67**:360–368.

Herrlich, P., Angel, P., Rahmsdorf, H. J., Mallick, U., Pöting, A., Hieber, L., Lücke-Huhle, C., and Schorpp, M., 1986, The mammalian genetic stress response, Adv. Enzyme Regul. **25**:485–504.

Herrlich, P., Ponta, H., and Rahmsdorf, H.J., 1992, DNA damage-induced gene expression: Signal transduction and relation to growth factor signaling, Rev. Physiol. Biochem. Pharmacol. **119**:187–216.

Hilgers, G., Clauss, I. M., Huez, G. A., and Rommelaere, J., 1991, Post-transcriptional effect of ultraviolet light on gene expression in human cells, Eur. J. Biochem. **201**:483–488.

Hill, C. K., Han, A., and Elkind, M. M., 1987, Promotion, dose rate, and repair processes in radiation-induced neoplastic transformation, Radiat. Res. **109**:347–351.

Hiscott, J., Marois, J., Garoufalis, J., and D'Addario, M., 1993, Characterization of a functional NF-κB site in the human interleukin 1 beta promoter: Evidence for a positive autoregulatory loop, Mol. Cell. Biol. **13**:6231–6240.

Holbrook, N. J., and Fornace, A. J., Jr., 1991, Response to adversity: Molecular control of gene activation following genotoxic stress, New Biol. **3**:825–833.

Hollander, C., and Fornace, A. J., Jr., 1989, Induction of fos RNA by DNA-damaging agents, Cancer Res. **49**:1687–1692.

Ijichi, A., Sakuma, S., and Tofilon, P. J., 1995, Hypoxia-induced vascular endothelial growth factor expression in normal rat astrocyte cultures, Glia **14**:87–93.

Jones, J. S., and Prakash, L., 1991, Transcript levels of the Saccharomyces cerevisiae DNA repair gene RAD18 increase in UV-irradiated cells and during meiosis but not during the mitotic cell cycle, Nucleic Acids Res. **19**:893–898.

Kaina, B., Stein, B., Schönthal, A., Rahmsdorf, H. J., Ponta, H., and Herrlich, P., 1989, An update of the mammalian UV response: Gene regulation and induction of a protective function, in: DNA Repair Mechanisms and Their Biological Implications in Mammalian Cells (M. W. Lambert and J. Laval, eds.), Plenum Press, New York, pp. 149–165.

Kartasova, T., and van de Putte, P., 1988, Cis- and trans-acting genetic elements responsible for induction of specific genes by tumor promoters, serum factors, and stress, in: Genes and Signal Transduction in Multistage Carcinogenesis (N. H. Colburn, ed.), Dekker, New York, pp. 415–440.

Kartasova, T., Cornelissen, B. J. C., Belt, P., and van de Putte, P., 1987, Effects of UV, 4-NQO, and TPA on gene expression in cultured human epidermal keratinocytes, Nucleic Acids Res. **15**:5945–5962.

Kasid, U., Pirollo, K., Dritschilo, A., and Chang, E., 1993, Oncogenic basis of radiation resistance, Adv. Cancer Res. **61**:195–233.

Kastan, M. B., Onyekwere, O., Sidransky, D., Vogelstein, B., and Craig, R. W., 1991, Participation of p53 protein in the cellular response to DNA damage, Cancer Res. **51**:6304–6311.

Kastan, M. B., Zhan, Q., El-Deiry, W. S., Carrier, F., Jacks, T., Walsh, W. V., Plunkett, B. S., Vogelstein, B., and Fornace, A. J., Jr., 1992, A mammalian cell cycle checkpoint pathway utilizing p53 and GADD45 is defective in ataxia telangiectasia, Cell **71**:587–597.

Kessis, T. D., Slebos, R. J., Nelson, W. G., Kastan, M. B., Plunkett, B. S., Han, S. M.,

Lorincz, A. T., Hedrick, L., and Cho, K. R., 1993, Human papillomavirus 16 E6 expression disrupts the p53-mediated cellular response to DNA damage, *Proc. Natl. Acad. Sci. USA* **90**:3988–3992.

Keys, S. M., and Tyrell, R. M., 1989, Heme oxygenase is the major 32-kDa stress protein induced in human skin fibroblasts by UVA radiation, hydrogen peroxide, and sodium arsenite, *Proc. Natl. Acad. Sci. USA* **86**:99–103.

Kharbanda, S., Ren, R., Pandey, P., Shafman, T. D., Feller, S. M., Weichselbaum, R. R., and Kufe, D. W., 1995, Activation of the c-Abl tyrosine kinase in the stress response to DNA-damaging agents, *Nature* **376**:785–788.

Komatsu, K., Sawada, S., Takeoka, S., Kodama, S., and Okumura, Y., 1993, Dose-rate effects of neutrons and γ-rays on the induction of mutation and oncogenic transformation in plateau-phase mouse m5S cells, *Int. J. Radiat. Biol.* **63**:469–474.

Koong, A. C., Chen, E. Y., and Giaccia, A. J., 1994, Hypoxia causes the activation of nuclear factor κB through the phosphorylation of IκBα on tyrosine residues, *Cancer Res.* **54**:1425–1430.

Kopp, E., and Ghosh, S., 1994, Inhibition of NF-κB by sodium salicylate and aspirin, *Science* **265**:956–959.

Krämer, M., Stein, B., Mai, S., Kunz, E., König, H., Loferer, H., Grunicke, H. H., Ponta, H., Herrlich, P., and Rahmsdorf, H. J., 1990, Radiation-induced activation of transcription factors in mammalian cells, *Radiat. Environ. Biophys.* **29**:303–313.

Krämer, M., Sachsenmaier, C., Herrlich, P., and Rahmsdorf, H. J., 1993, UV-irradiation induced interleukin 1 and basic fibroblast growth factor synthesis and release mediate part of the UV response, *J. Biol. Chem.* **268**:6734–6741.

Kretzchmar, M., Meisterenst, M., Scheidereit, C., Li, G., and Roeder, R. G., 1992, Transcriptional regulation of the HIV-1 promoter by NF-κB *in vitro*, *Genes Dev.* **6**:761–774.

Kunsch, C., and Rosen, C. A., 1993, NF-κB subunit-specific regulation of the interleukin-8 promoter, *Mol. Cell. Biol.* **13**:6137–6146.

Lacoste, J., D'Addario, M., Roulston, A., Wainberg, M. A., and Hiscott, J., 1990, Cell-specific differences in activation of NF-κB regulatory elements of human immunodeficiency virus and β interferon promoters by tumor necrosis factor, *J. Virol.* **64**:4726–4734.

Lambert, M. E., Ronai, Z. A., Weinstein, I. B., and Garrels, J. I., 1989, Enhancement of major histocompatibility class I protein synthesis by DNA damage in cultured human fibroblasts and keratinocytes, *Mol. Cell. Biol.* **9**:847–850.

Liang, P., and Pardee, A. B., 1992, Differential display of eukaryotic messenger RNA by means of polymerase chain reaction, *Science* **257**:969–971.

Liang, P., Averboukh, L., Keyomarsi, K., Sager, R., and Pardee, A. B., 1992, Differential display and cloning of messenger RNAs from human breast cancer versus mammary epithelial cells, *Cancer Res.* **52**:6966–6968.

Liang, P., Averboukh, L., and Pardee, A. B., 1993, Distribution and cloning of eukaryotic mRNAs by means of differential display: Refinements and optimization, *Nucleic Acids Res.* **21**:3269–3275.

Libertin, C. R., Panozzo, J., Groh, K. R., Chang-Liu, C.-M., Schreck, S., and Woloschak, G. E., 1994, Effects of gamma rays, ultraviolet radiation, sunlight, microwaves, and electromagnetic fields on gene expression mediated by human immunodeficiency virus promoter, *Radiat. Res.* **140**:91–96.

Lieberman, M. W., Beach, L. R., and Palmiter, R. D., 1983, Ultraviolet radiation-induced metallothionein-I gene activation is associated with extensive DNA demethylation, *Cell* **35**:207–214.

Lin, C. S., Goldthwait, D. A. and Samols, D., 1990, Induction of transcription from the long terminal repeat of Moloney murine sarcoma provirus by UV-irradiation, X-irradiation, and phorbol ester, *Proc. Natl. Acad. Sci. USA* **87**:36–40.

Liou, H. C., and Baltimore, D., 1993, Regulation of the NF-κB/rel transcription factor and IκB inhibitor system, *Curr. Opin. Cell. Biol.* **5**:477–487.

Mai, S., Stein, B., Vandenberg, S., Kaina, B., Lücke-Huhle, C., Ponta, H., Rahmsdorf, H. J., Kraemer, M., Gebel, S., and Herrlich, P., 1989, Mechanisms of the UV response in mammalian cells, *J. Cell Sci.* **94**:609–615.

Maltzman, W., and Czyzyk, L., 1984, UV irradiation stimulates levels of p53 cellular tumor antigen in nontransformed mouse cells, *Mol. Cell. Biol.* **4**:1689–1694.

Mangan, D. F., Robertson, B., and Wahl, S. M., 1992, IL-4 enhances programmed cell death (apoptosis) in stimulated human monocytes, *J. Immunol.* **148**:1812–1816.

Martin, M., Cefaix, J.-L., Pinton, P., Crechet, F., and Daburton, F., 1993, Temporal modulation of TGR-β1 and β-actin gene expression in pig skin and muscular fibrosis after ionizing radiation, *Radiat. Res.* **134**:63–70.

Matsui, M. S., and DeLeo, V. A., 1990, Induction of protein kinase C activity by ultraviolet radiation, *Carcinogenesis* **11**:229–234.

McAnulty, R. J., Moores, S. R., Talbot, R. J., Bishop, J. E., Mays, P. K., and Laurent G. J., 1991, Long-term changes in mouse lung following inhalation of a fibrosis-inducing dose of ^{239}Pu O$_2$: Changes in collagen synthesis and degradation rates, *Int. J. Radiat. Biol.* **59**:229–238.

McKenna, W., Iliakis, G., Weiss, M. C., Bernhard, E. J., and Muschel, R. J., 1991, Increased G2 delay in radiation-resistant cells obtained by transformation of primary rat embryo cells with oncogenes H-*ras* and v-*myc*, *Radiat. Res.* **125**:283–287.

McWilliams, R. S., Gross, W. G., Kaplan, J. G., and Birnboim, H. C., 1983, Rapid rejoining of DNA strand breaks in resting human lymphocytes after irradiation by low doses of ^{60}Co γ-rays or 14.6-MeV neutrons, *Radiat. Res.* **94**:499–507.

Messer, G., Weiss, E. H., and Baeuerle, P. A., 1990, Tumor necrosis factor beta (TNF-β) induces binding of the NF-κB transcription factor to a high-affinity κB element in the TNF-β promoter, *Cytokine* **2**:389–397.

Miskin, R., and Ben-Ishai, R., 1981, Induction of plasminogen activator by UV light in normal and xeroderma pigmentosum fibroblasts, *Proc. Natl. Acad. Sci. USA* **78**:6236–6240.

Mohan, N., and Meltz, M. L., 1994, Induction of nuclear factor κB after low-dose ionizing radiation involves a reactive oxygen intermediate signaling pathway, *Radiat. Res.* **140**:97–104.

Morrey, J. D., Bourn, S. M., Bunch, T. D., Jackson, M. K., Sidwell, R. W., Barrows, L. R., Daynes, R. A., and Rosen, C. A., 1991, In vivo activation of human immunodeficiency virus type 1 long terminal repeat by UV type A (UVA) light plus psoralen and UVB light in the skin of transgenic mice, *J. Virol.* **65**:5045–5051.

Mossman, T. R., and Coffman, R. L., 1989, TH1 and TH2 cells: Different patterns of lymphokine secretion lead to different functional properties, *Ann. Rev. Immunol.* **7**:145–174.

Munson, G., and Woloschak, G. E., 1990, Differential effect of ionizing radiation on transcription in repair-deficient and repair-proficient mice, *Cancer Res.* **50:** 5045–5048.

Muschel, R. J., Zhang, H. B., and McKenna, W. B., 1993, Differential effect of ionizing radiation on the expression of cyclin A and cyclin B in HeLa cells, *Cancer Res.* **53:**1128–1135.

Nelson, W. G., and Kastan, M. B., 1994, DNA strand breaks: The DNA template alterations that trigger p53-dependent DNA damage response pathways, *Mol. Cell. Biol.* **14:**1815–1823.

Neta, R., and Oppenheim, J. J., 1988, Cytokines in therapy of radiation injury, *Blood* **72:**1093–1095.

Neta, R., and Oppenheim, J. J., 1991, Radioprotection with cytokines. Learning from nature to cope with radiation damage, *Cancer Cell* **3:**391–396.

Neta, R., Oppenheim, J. J., and Douches, S. D., 1988, Interdependence of the radioprotective effects of human recombinant IL-1, TNF, G-CSF, and murine recombinant G-CSF, *J. Immunol.* **140:**108–111.

Neta, R., Monroy, R., and MacVittie, T. J., 1989, Utility of interleukin 1 in therapy of radiation injury as studied in small and large animal models, *Biotherapy* **1:** 301–311.

Oleinick, N. L., and Evans, H. H., 1985, Poly(ADP-ribose) and the response of cells to ionizing radiation, *Radiat. Res.* **101:**29–46.

Oliff, A., Defeo-Jones, D., Boyer, M., Martinez, D., Diefer, D., Vuocolo, G., Wolfe, A., and Socher, S. H., 1987, Tumors secreting human TNF/cachexin induce cachexia in mice, *Cell* **50:**555–563.

Ossanna, N., Peterson, K. R., and Mount, D. W., 1987, UV-inducible SOS response in *Escherichia coli, Photobiology* **45:**905–908.

Owens, G. P., Hahn, W. E., and Cohen, J. J., 1991, Identification of mRNAs associated with programmed cell death in immature thymocytes, *Mol. Cell. Biol.* **8:**4177–4188.

Painter, R. B., and Young, B. R., 1987, DNA synthesis in irradiated mammalian cells, *J. Cell Sci. Suppl.* **6:**207–214.

Panozzo, J., Bertoncini, D., Miller, D., Libertin, C. R., and Woloschak, G. E., 1991, Modulation of expression of virus-like elements following exposure of mice to high-and low-LET radiations, *Carcinogenesis* **12:**801–804.

Papathanasiou, M. A., Kerr, N. C. K., Robbins, J. H., McBride, O. W., Alamo, I., Jr., Barrett, S. F., Hickson, I. D., and Fornace, A. J., Jr., 1991, Induction by ionizing radiation of the gadd45 gene in cultured human cells: Lack of mediation by protein kinase C, *Mol. Cell. Biol.* **11:**1009–1016.

Peak, J. G., Woloschak, G. E., and Peak, M. J., 1991, Enhanced expression of protein kinase C gene caused by solar radiation, *Photochem. Photobiol.* **53:** 395–397.

Perkins, N. D., Edwards, N. L., Duckett, C. S., and Agranoff, A. B., 1993, A cooperative interaction between NF-κB and Sp1 is required for HIV-1 enhancer activation, *EMBO J.* **12:**3551–3558.

Peter, R. U., Beetz, A., Ried, C., Michel, G., van Beuningen, D., and Ruzicka, T., 1993, Increased expression of the epidermal growth factor receptor in human epidermal keratinocytes after exposure to ionizing radiation, *Radiat. Res.* **136:** 65–70.

Prasad, A. V., Mohan, N., Chandrasekar, B., and Meltz, M. L., 1994, Activation of nuclear factor κB in human lymphoblastoid cells by low-dose ionizing radiation, *Radiat. Res.* **138**:367–372.

Ramsamooj, P., Kasid, U., and Dritschilo, A., 1992, Differential expression of proteins in radioresistant and radiosensitive human squamous carcinoma cells, *J. Natl. Cancer Inst.* **84**:622–628.

Rand, A., and Koval, T. M., 1994, Coordinate regulation of proteins associated with radiation resistance in cultured insect cells, *Radiat. Res.* **138**:S13–S16.

Rao, J. S., Steck, P. A., Tofilon, P., Boyd, D., Ali-Osman, F., Stetler-Stevenson, W. G., Liotta, L. A., and Sawaya, R., 1993, Role of plasminogen activator and of 92-kDa type IV collagenase in glioblastoma invasion using an *in vitro* matrigel model, *J. Neurooncol.* **18**:129–138.

Rao, J. S., Rayford, A., Yamamoto, M., Ang, K. K., Tofilon, P., and Sawaya, R., 1994, Modulation of fibrinolysis by ionizing radiation, *J. Neurooncol.* **22**:161–171.

Remy, J., Wegrowski, J., Crechet, F., Martin, M., and Daburon, F., 1991, Long-term overproduction of collagen in radiation-induced fibrosis, *Radiat. Res.* **125**:14–19.

Ronai, Z. A., and Weinstein, I. B., 1990, Identification of ultraviolet-inducible proteins that bind to a TGACAACA sequence in the polyoma virus regulatory region, *Cancer Res.* **50**:5374–5381.

Ronai, Z. A., Okin, E., and Weinstein, I. B., 1988, Ultraviolet light induces expression of oncogenes in rat fibroblasts and human keratinocyte cells, *Oncogene* **2**:201–204.

Ronai, Z. A., Lambert, M. E., and Weinstein, I. B., 1990, Inducible cellular responses to ultraviolet light irradiation and other mediators of DNA-damage in mammalian cells, *Cell Biol. Toxicol.* **6**:105–126.

Ronai, Z. A., Rutberg, S., and Yang, Y. M., 1994, UV-response element (TGACAACA) from rat fibroblasts to human melanomas, *Environ. Mol. Mutagen.* **23**:157–163.

Rosen, C. F., Gajic, D., and Drucker, D. J., 1990, UV radiation induction of ornithine decarboxylase in rat keratinocytes, *Cancer Res.* **50**:2631–2635.

Rotem, N., Axelrod, J. H., and Miskin, R., 1987, Induction of urokinase-type plasminogen activator by UV light in human fetal fibroblasts is mediated through a UV-induced secreted protein, *Mol. Cell. Biol.* **7**:622–631.

Rozek, D., and Pfeifen, G. P., 1993, *In vivo* protein–DNA interactions at the *c-jun* promoter: Preformed complexes mediate the UV response, *Mol. Cell. Biol.* **13**:5490–5499.

Rutberg, S. E., Yang, Y. M., and Ronai, Z., 1992, Functional role of the ultraviolet light responsive element (URE: TGACAACA) in the transcription and replication of polyoma DNA, *Nucleic Acids Res.* **20**:4305–4310.

Sahijdak, W. M., Yang, C.-R., Zuckerman, J. S., Meyers, M., and Boothman, D. A., 1994, Alterations in transcription factor binding in radioresistant human melanoma cells after ionizing radiation, *Radiat. Res.* **138**:47–51.

Sakakeeny, M. A., Harrington, M., Leif, J., Merrill, W., Pratt, D., Romanik, E., McKenna, M., Fitzgerald, T. J., and Greenberger, J. S., 1994, Effects of gamma-irradiation on the M-CSF-promoter linked to a chloramphenicol acetyl transferase reporter gene expressed in a clonal murine bone marrow stromal cell line, *Stem Cells* **12**:87–94.

Sakuma, S., Saya, H., Ijichi, A., and Tofilon, P. J., 1995, Radiation induction of the receptor tyrosine kinase gene Ptk-3 in normal rat astrocytes, *Radiat. Res.* **143**:1–7.

Sawaya, R., Tofilon, P. J., Mohanam, S., Ali-Osman, F., Liotta, L. A., Stetler-Stevenson, W. G., and Rao, J. S., 1994, Induction of tissue-type plasminogen activator and 72-kDa type-IV collagenase by ionizing radiation in rat astrocytes, *Int. J. Cancer* **56:**214–218.

Schmid, R. M., Perkins, N. D., Duckett, C. S., Andrews, P. C., and Nabel, G. J., 1991, Cloning of an NF-κB subunit which stimulates HIV transcription in synergy with p65, *Nature* **352:**733–736.

Schorpp, M., Mallick, U., Rahmsdorf, H. J., and Herrlich, P., 1984, UV-induced extracellular factor from human fibroblasts communicates the UV response to nonirradiated cells, *Cell* **37:**861–868.

Schreck, S., Rieber, P., and Baeuerle, P. A., 1991, Reactive oxygen intermediates as apparently widely used messengers in the activation of the NF-κB transcription factor and HIV-1, *EMBO J.* **10:**2247–2258.

Schreck, S., Panozzo, J., Milton, J., Libertin, C. R., and Woloschak, G. E., 1995, The effects of multiple UV exposures on HIV-LTR expression, *Photochem. Photobiol.* **61:**378–382.

Sellins, K. S., and Cohen, J. J., 1987, Gene induction by γ-irradiation leads to DNA fragmentation in lymphocytes, *J. Immunol.* **139:**3199–3206.

Servomaa, K., and Rytömaa, T., 1990, UV light and ionizing radiations cause programmed death of rat chloroleukaemia cells by inducing retropositions of a mobile DNA element (L1Rn), *Int. J. Radiat. Biol.* **57:**331–343.

Shadley, J. D., Afzal, V., and Wolff, S., 1987, Characterization of the adaptive response to ionizing radiation induced by low doses of X-rays to human lymphocytes, *Radiat. Res.* **111:**511–517.

Sidjanin, D., Grdina, D., and Woloschak, G. E., 1996, UV-induced changes in cell cycle and gene expression within rabbit lens epithelial cells, *Photochem. Photobiol.* **63:**79–85.

Siebert, P. D., and Fukuda, M., 1985, Induction of cytoskeletal vimentin and actin gene expression by a tumor-promoting phorbol ester in the human leukemic cell line K562, *J. Biol. Chem.* **260:**3868–3874.

Simon, M. M., Aragane, Y., Schwarz, A., Luger, T. A., and Schwarz, T., 1994, UVB light induces nuclear factor κB (NFκB) activity independently from chromosomal DNA damage in cell-free cytosolic extracts, *J. Invest. Dermatol.* **102:** 422–427.

Singh, S. P., and Lavin, M. F., 1989, DNA-binding protein activated by gamma radiation in human cells, *Mol. Cell. Biol.* **10:**5279–5285.

Stanley, S. K., Folks, T. M., and Fauci, A. S., 1989, Induction of expression of the human immunodeficiency virus in a chronically infected promonocytic cell line by ultraviolet irradiation, *AIDS Res. Hum. Retrovir.* **5:**375–384.

Stein, B., Rahmsdorf, H. J., Schönthal, A., Büscher, M., Ponta, H., and Herrlich, P., 1988, The UV induced signal transduction pathway to specific genes, in: *Mechanisms and Consequences of DNA Damage Processing* (E. Friedberg and P. Hanawalt, eds.), Liss, New York, pp. 557–570.

Stein, B., Kramer, M., Rahmsdorf, H. J., Ponta, H., and Herrlich, P., 1989a, UV-induced transcription from the HIV-1 LTR and UV-induced secretion of an extracellular factor that induces HIV-1 transcription in non-irradiated cells, *J. Virol.* **63:**4540–4544.

Stein, B., Rahmsdorf, H. J., Steffen, A., Litfin, M., and Herrlich, P., 1989b, UV-induced DNA damage is an intermediate step in UV-induced expression of human immunodeficiency virus type I, collagenase, c-*fos*, and metallothionein, *Mol. Cell. Biol.* **9:**5169–5181.

Stein, B., Angel, P., van Dam, H., Ponta, H., Herrlich, P., van der Eb, A., and Rahmsdorf, H. J., 1992, UV-induced c-jun gene transcription: Two AP-1 like binding sites acting additively mediate the response, *Photochem. Photobiol.* **55:** 409–415.

Sun, X., Shimizu, H., and Yamamoto, K., 1995, Identification of a novel p53 promoter element in genotoxic stress-inducible p53 gene expression, *Mol. Cell. Biol.* **8:**4489–4496.

Uchiume, T., Kohno, K., Tanimura, H., Matsu, K., Sato, S., Uchida, Y., and Kuwano, M., 1993, Enhanced expression of the human multidrug resistance 1 gene in response to UV light irradiation, *Cell Growth Differ.* **4:**147–157.

Uckun, F. M., Tuel-Ahlgren, L. M., Song, C. W., Waddick, K., Myers, D. E., Kirihara, J., Ledbetter, J. A., and Schieven, G. L., 1992, Ionizing radiation stimulates unidentified tyrosine-specific protein kinases in human B-lymphocyte precursors, triggering apoptosis and clonogenic cell death, *Proc. Natl. Acad. Sci. USA* **89:**9005–9009.

Uckun, F. M., Schieven, G. L., Tuel-Ahlgren, L. M., Dibirdik, I., Myers, D. E., Ledbetter, J. A., and Song, C. W., 1993, Tyrosine phosphorylation is a mandatory proximal step in radiation-induced activation of the protein kinase C signaling pathway in human B-lymphocyte precursors, *Proc. Natl. Acad. Sci. USA* **90:**252–256.

Valerie, K., and Rosenberg, H., 1990, Chromatin structure implicated in activation of HIV-1 gene expression by ultraviolet light, *New Biol.* **2:**712–718.

Valerie, K., Delers, A., Bruck, C., Thiriart, C., Rosenberg, H., Debouck, C., and Rosenberg, M., 1988, Activation of human immunodeficiency virus type I by DNA damage in human cells, *Nature* **333:**78–81.

Vandenberg, S., Kaina, B., Rahmsdorf, H. J., Ponta, H., and Herrlich, P., 1991, Involvement of Fos in spontaneous and ultraviolet light-induced genetic changes, *Mol. Carcinogen* **4:**460–466.

van der Schans, G. P., Paterson, M. C., and Gross, W. G., 1983, DNA strand break and rejoining in cultured human fibroblasts exposed to fast neutrons or gamma rays, *Int. J. Radiat. Biol.* **44:**75–85.

Vanetti, M., 1988, Der unterschiedliche Beitrag der Motive in SV40 Enhancer zur Induktion mit UV Strahlung, Diplomarbeit, Universitat Karlsruhe.

Vrdoljak, E., Borchardt, P. E., Bill, C. A., Stephens, L. C., and Tofilon, P. J., 1994, Influence of x-rays on early response gene expression in rat astrocytes and brain tumor cell lines, *Int. J. Radiat. Biol.* **66:**739–746.

Wade, M. H., and Trosko, J. E., 1983, Enhanced survival and decreased mutation frequency after photoreactivation of UV damage in rat kangaroo cells, *Mutat. Res.* **112:**231–243.

Warner, H. R., and Price, A. R., 1989, Involvement of DNA repair in cancer and aging, *J. Gerontol.* **44:**45–54.

Weber, K. J., Schneider, E., Kiefer, J., and Kraft, G., 1990, Heavy ion effects on yeast: Inhibition of ribosomal RNA synthesis, *Radiat. Res.* **123:**61–67.

Weichselbaum, R. R., Hallahan, D., Fuks, Z., and Kufe, D., 1994, Radiation induction of immediate early genes: Effectors of the radiation-stress response, *Int. J. Radiat. Oncol. Biol. Phys.* **30:**229–234.

Woloschak G. E., and Chang-Liu, C.-M., 1990, Differential modulation of specific gene expression following high- and low-LET radiations, *Radiat. Res.* **124:** 183–187.

Woloschak G. E., and Chang-Liu, C.-M., 1991, Expression of cytoskeletal elements in proliferating cells following radiation exposure, *Int. J. Radiat. Biol.* **59:**1173–1183.

Woloschak, G. E., and Chang-Liu, C.-M., 1992, Effects of low-dose radiation on gene expression in Syrian hamster embryo cells: Comparison of JANUS neutrons and gamma rays, in: *Low Dose Irradiation and Biological Defense Mechanisms* (T. Sugahara, L. A. Sagan, and T. Aoyama, eds.), Elsevier, Amsterdam, pp. 239–242.

Woloschak, G. E., and Chang-Liu, C.-M., 1995, Modulation of expression of genes encoding nuclear proteins following exposure to JANUS neutrons or γ-rays, *Cancer Lett.* **97:**169–175.

Woloschak, G. E., Liu, C.-M., and Shearin-Jones, P., 1990a, Regulation of protein kinase C by ionizing radiation, *Cancer Res.* **50:**3963–3967.

Woloschak, G. E., Liu, C.-M., Jones, P. S., and Jones, C. A., 1990b, Modulation of gene expression in Syrian hamster embryo cells following ionizing radiation, *Cancer Res.* **50:**339–344.

Woloschak, G. E., Shearin-Jones, P., and Chang-Liu, C.-M., 1990c, Effects of ionizing radiation on expression of genes encoding cytoskeletal elements: Kinetics and dose effects, *Mol. Carcinogen.* **3:**374–378.

Woloschak, G. E., Churchill, M. E., and Libertin, C. R., 1991, Immunological disorders characterizing the "wasted" mouse: A review, *Immunol. (Life Sci. Adv.)* **10:**95–104.

Woloschak, G. E., Chang-Liu, C.-M., Panozzo, J., and Libertin, C. R., 1994, Low doses of neutrons induce changes in gene expression, *Radiat. Res.* **138:**S56–S59.

Woloschak, G. E., Paunesku, T., Chang-Liu, C.-M., and Grdina, D. J., 1995a, Expression of thymidine kinase messenger RNA and a related transcript is modulated by radioprotector WR1065, *Cancer Res.* **55:**4788–4792.

Woloschak, G. E., Felcher, P., and Chang-Liu, C.-M. 1995b, Combined effects of ionizing radiation and cycloheximide on gene expression, *Mol. Carcinogen.* **13:** 44–49.

Woloschak, G. E., Felcher, P., and Chang-Liu, C.-M., 1995c, Expression of cytoskeletal and matrix genes following exposure to ionizing radiation: Dose-rate effects and protein synthesis requirements, *Cancer Lett.* **92:**135–141.

Woloschak, G. E., Panozzo, J., Schreck, S., and Libertin, C. R., 1995d, Salicylic acid inhibits ultraviolet- and *cis*-platinum-induced human immunodeficiency virus expression, *Cancer Res.* **55:**1696–1700.

Woloschak, G. E., Chang-Liu, C.-M., Chung, J., and Libertin, C. R., 1996, Expression of enhanced spontaneous and γ-ray-induced apoptosis by lymphocytes of the wasted mouse, *Int. J. Radiat. Biol.* **69:**47–55.

Yang, Y. M., Rutberg, S. E., Foiles, P. G., and Ronai, Z., 1993, Expression patterns of proteins that bind to the ultraviolet-response elements (TGACAACA) in human keratinocytes, *Mol. Carcinogen.* **7:**36–43.

Yonish-Rouach, E., Resnitzky, D., Lotem, J., Sachs, L., Kimichi, A., and Oren, M., 1991, Wild-type p53 induces apoptosis of myeloid leukaemic cells that is inhibited by interleukin-6, *Nature* **352:**345–347.

Zhan, O., Lord, K. A., Alamo, I., Jr., Hollander, M. C., Carrier, F., Ron, D., Kohn, K. W., Hoffman, B., Liebermann, D. A., and Fornace, A. J., Jr., 1994, The *gadd* and *MyD* genes define a novel set of mammalian genes encoding acidic proteins that synergistically suppress cell growth, *Mol. Cell Biol.* **14:**2361–2371.

Zmudzka, B., and Beer, J. Z., 1990, Yearly review: Activation of human immunodeficiency virus by UV radiation, *Photochem. Photobiol.* **52:**1153–1162.

Zubiaga, A. M., Mūnoz, E., and Huber, B., 1992, IL-4 and IL-2 selectively rescue T cell subsets from glucocorticoid-induced apoptosis, *J. Immunol.* **146:**3857–3863.

9

Adaptation to Ionizing Radiation in Mammalian Cells

R. E. J. MITCHEL, E. I. AZZAM,
and S. M. DE TOLEDO

1. INTRODUCTION

The concept of an ionizing radiation-induced increase in resistance against the effects of a subsequent exposure is an accepted and reasonably well-understood process in both prokaryotes (Walker, 1984, 1985) and nonmammalian eukaryotes (Calkins, 1967; Boreham *et al.*, 1990, 1991; Boreham and Mitchel, 1991, 1993, 1994; Mitchel and Morrison, 1982, 1984, 1987; Koval, 1986, 1988). Such processes however, have been more difficult to demonstrate in mammalian cells, where their existence and/or significance have been very controversial (Olivieri *et al.*, 1984; Olivieri and Bosi, 1990; Wiencke *et al.*, 1987, Wilson, 1989; Wojcik *et al.*, 1992a,b; Wolff, 1992 a,b). This lack of general acceptance, while reflecting the lack of volume of the data as well as the variability

R. E. J. MITCHEL, E. I. AZZAM, and S. M. DE TOLEDO • Radiation Biology and Health Physics Branch, Atomic Energy of Canada Limited, Chalk River Laboratories, Chalk River, Ontario K0J 1J0, Canada.

Stress-Inducible Processes in Higher Eukaryotic Cells, edited by Koval. Plenum Press, New York, 1997.

221

noted above and a lack of direct evidence of an influence on whole animal risk, also reflects the fact that the concept challenges long entrenched and widely held beliefs and practices, both scientific and public, on which all radiation protection programs and cancer risk estimates are based. Consequently, demonstration of the existence and quantification of the significance of adaptation to radiation in mammalian cells has, potentially, large social and economic implications.

2. ADAPTATION TO RADIATION IN HUMAN SKIN FIBROBLASTS

The early work of Wolff and co-workers (Olivieri *et al.*, 1984; Wiencke *et al.*, 1986; Wolff *et al.*, 1988, 1989) demonstrated that human lymphocytes could adapt after a low dose of ionizing radiation, and become less sensitive to the chromosomal damaging effects of a subsequent larger dose. These experiments were important because they demonstrated that the adaptive response to radiation, known to exist in prokaryotes and nonmammalian eukaryotes, had in fact been evolutionarily conserved in humans. However, the significance of this finding, in terms of cancer risk from radiation, was not clear because peripheral blood lymphocytes are normally nondividing cells and are not considered a cell type at risk in humans. Therefore, primary human skin fibroblasts (AG1522) are discussed in this section as a representative dividing human cell type, potentially at risk for transformation from radiation exposure, to investigate the characteristics of adaptation to radiation.

2.1. Survival

Historically, the endpoint most frequently used to detect adaptation to radiation in prokaryotes and nonmammalian eukaryotes has been survival. In mammalian cells, however, survival changes are difficult to detect and the magnitude of the change induced is very small relative to that seen in nonmammalian cells. Our results showing a radiation-induced increase in survival to a subsequent killing dose were the first clear indication that this effect could also be demonstrated in human cells (Azzam *et al.*, 1992). Chronic, low-dose-rate (0.0024 Gy/min) exposures of fibroblasts (incubating at 37°C) were used to stimulate the adaptive response, and a large acute exposure was used as the test dose. The maximum increase in survival detected was only about two-fold (Fig. 1), using a standard colony forming assay. This can be compared with the orders of magnitude change seen, for example, in yeast (Mitchel and Morrison, 1984). It is interesting that at these low-dose-rate exposures used to adapt the human cells, adaptation, in terms of survival to a second acute exposure, could be seen after a total adapting dose of 2–5 Gy. These are large doses, far in excess of those normally of concern in occupational radiation protection, but they are similar to doses used in cancer radiotherapy. This then raises the question of radiation-induced radioresistance in fractionated radiotherapy protocols for cancer treatment, where tumor cell survival is the critical endpoint for therapy success. Other groups (Marples and Joiner, 1993; Marples et al., 1994; Wouters and Skarsgard, 1994; Lambin et al., 1993, 1994) have undertaken detailed examinations of survival curves in human and other mammalian cells. The multiphasic nature of these curves in the low-dose region (typically below 1 Gy) has also been interpreted as evidence for induced radioresistance resulting in increased survival as dose increased. To our knowledge, however,

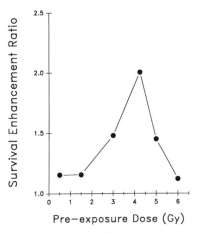

FIGURE 1. Chronic (0.0024 Gy/min) radiation preexposure enhancement in the survival of human fibroblasts exposed to an acute 4.25-Gy dose.

the question of radioresistance induced by fractionated radiotherapy of human tumor cells has not been pursued.

Another measure of survival as an indicator of induced radioresistance is colony size. An assessment was made of the number
of cells in colonies originating from cells that survived an acute
radiation exposure, and that had or had not received a previous
chronic adapting exposure (Azzam et al., 1992). A large chronic
exposure of 4.25 Gy preceding a similarly large acute exposure
(total 8.5 Gy) resulted in twice as many colonies. In addition, when
the colonies were examined 7 days after plating cells for survival,
the colonies had about fourfold as many cells per colony, compared with the acute 4.25-Gy exposure alone (Fig. 2). A recent
automated analysis of colony size versus dose showed similar
effects (Spadinger et al., 1994). These results are rather disturbing
for cancer radiotherapy, as they suggest that if tumor cells respond in a fashion similar to these normal fibroblasts, cells surviving
one fraction of a fractionated protocol may regrow more quickly
and increase the rate of repopulation of the tumor site. They also
suggest that about 1 week after the end of exposure, a tumor that
received two dose fractions at a somewhat low dose rate could have
about one order of magnitude more viable cells than if it had
received only half that dose at an acute dose rate. It is obviously
important to understand the effect of dose rate on this adaptive

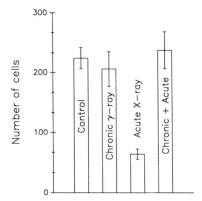

FIGURE 2. The influence on the number of cells per colony of a chronic (0.0025
Gy/min) 4.25-Gy γ-ray dose in human fibroblasts surviving a subsequent 4.25-Gy
acute x-ray exposure.

process for cell survival as it may contribute to radiotherapy failures in a fractionated protocol.

2.2. Adaptation, Chromosomal Breaks, Dose, and Dose Rate

Using the human skin fibroblasts, various characteristics of the adaptive response have been investigated by using the endpoint of micronucleus (MN) formation as a measure of unrepaired chromosomal breaks. To reduce the additional effects of possible variations with cell cycle position, we have routinely investigated adaptation in confluence-arrested cells, with about 90% in G_0/G_1 as measured by flow cytometry. The frequency of MN formation following various radiation adapting and test doses (all given to confluence-arrested cells) was assessed in binucleate cells (BNC) arrested at the first subsequent cell division by replating the cells at exponential phase density in the presence of cytochalasin B, a compound that allows nuclear but not cytoplasmic division.

Using the MN assay, it was demonstrated that confluence-arrested human fibroblast cells exposed to 0.5 Gy at a low dose rate (0.0025 Gy/min) and subsequently exposed to an acute 4-Gy test dose (while still confluence arrested) developed less MN/BNC than cells exposed to the acute dose alone (Fig. 3A), indicating that adapted cells are better protected against DNA double-strand breaks that lead to chromosomal breaks and MN. Figure 3A also shows that introducing a 5-hr incubation at 37°C between the 0.5-Gy adapting dose and the 4-Gy test dose allowed greater expression of the adaptation process, further reducing the MN frequency resulting from the test dose.

Figure 3B shows the results of the same experiment when the initial 0.5-Gy adapting dose was delivered at high dose rate (2 Gy/min). Given sequentially, without incubation between the doses, the effects were additive as expected. As the incubation period between the two doses increased, the overall effect became less than that of the acute 4-Gy dose alone, indicating that some adaptation occurred even after an acute dose. However, even after 15 hr of incubation the extent of the reduction in MN frequency was not as great as after the identical dose given at a low dose rate (Fig. 3A) indicating that low dose rates are more effective than acute dose rates at inducing the adaptive response in human cells. An earlier study using human lymphocytes also showed dose rate

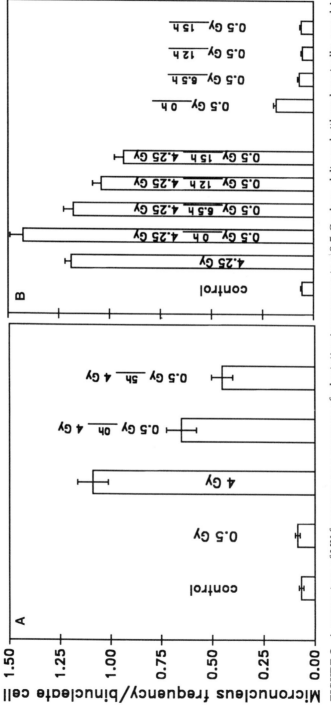

FIGURE 3. A comparison of MN frequency as a measure of adaptation in response to a 0.5-Gy dose, delivered either chronically, panel A (0.0025 Gy/min), or acutely, panel B (2 Gy/min), to human fibroblasts subsequently exposed to an acute (2 Gy/min) test dose of about 4 Gy.

effects for induction of the adaptive response, but in that study high dose rates appeared ineffective (Shadley and Wiencke, 1989).

When delivered chronically, the adapting dose of radiation can vary widely in total dose and still be effective. Doses of 4.25 Gy at 0.003 Gy/min and 0.5 Gy at 0.0005 Gy/min both produced about the same overall reduction in MN frequency (compare Azzam *et al.*, 1992, and Fig. 3A). Flow cytometric analysis indicated that these results were not related to changes in cell cycle distribution (Azzam *et al.*, 1992). Also tested was the possibility that induction of protective processes leading to a reduced frequency of MN formation occurred only in a subset of the exposed cell population. By examining changes in the number of MN formed in BNC, a comparison was made between adapted and unadapted cells, subsequently exposed to a large acute dose. It was observed that in the adapted cells there was a general downward shift in the number of MN observed in any one BNC. This was seen as an increase in the number of BNC with no MN, but also as a decrease in the number of BNC with two MN and an increase in the number with one MN (Azzam *et al.*, 1994b). These results indicate that even in those cells that were unlikely to survive the exposure (i.e., had one or more broken chromosomes remaining at the time of cell division), repair of DNA double-strand breaks was enhanced such that the number of broken chromosomes was reduced.

2.3. Induction of Adaptation by Heat

It is clear that exposure of human cells to ionizing radiation induces an adaptive response, increasing the resistance of these cells to the effects of further ionizing radiation exposure. However, some preliminary data indicate that a different kind of stress can elicit a similar response (Boreham *et al.*, 1994). A mild heat exposure (40°C) for a short time (15 min) given 12 hr prior to a test 4-Gy acute radiation exposure significantly reduced the frequency of radiation-induced MN (52.2% to 43.5%, $p < 0.024$) in human skin fibroblasts. As this mild hyperthermia is unlikely to introduce DNA damage, in contrast to ionizing radiation used as an adapting agent, the result indicates that adaptation to radiation can be induced by a non-DNA-damaging stress. As the hyperthermia treatment used was well within the range of common human fever, the result also suggests that the DNA repair capability and the risk of chromosomal deletions from a radiation exposure are not constant, and can be significantly altered not only by prior radiation exposure, but also by the common stress of fever.

2.4. Gene Expression

The treatment of normal human fibroblasts with 3.6 Gy of ^{60}Co γ rays, delivered at low dose rate, affects the transcript levels of some genes (de Toledo et al., 1995). An increase (twofold) was observed for *gadd45* and decreased levels were observed for the genes *cyclin A* (fourfold), *cyclin B* (twofold), *topoisomerase IIα* (fourfold), and *facl* (threefold). No changes were observed in the transcript levels for the genes β-actin, β-microglobulin, sod-1, uracil DNA glycosylase, topoisomerase I and IIβ, and cyclins C, D1, D2, D3, and E.

The cyclins are proteins whose levels vary throughout the cell cycle. They are responsible for conferring activity on their associated partners, protein kinases (cdks), at the appropriate time during the cell cycle. Cyclin A plays a role in the execution of S phase as well as during the G_2–M phase transition (Pagano et al., 1992). Cyclin B is involved in the control of the G_2–M transition (reviewed in Xiong and Beach, 1991). In mammalian cells three different classes of cyclins (C, D, and E) are expressed in the G_1 phase (reviewed in Sherr, 1993). The role of cyclin C is not clear but cyclins D1 and E seem to be involved in the G_1–S phase transition (Resnitzky et al., 1994). Ionizing radiation treatments lead to cell cycle delays in G_1 and G_2 phases, and decreased levels of cyclin A protein and cyclin B protein and mRNA have been observed in G_1- and in G_2-arrested cells respectively (Dulic et al., 1994; Mushel et al., 1991). The decreased levels of *cyclin A* and *cyclin B* transcripts that were observed in plateau phase AG1522 cells after irradiation may indicate that even in quiescent cells, ionizing radiation is able to induce a process whereby further downregulation of these cyclin transcript levels is achieved, probably leading to or as a consequence of repositioning of the cells at a different stage in G_0–G_1 (de Toledo et al., 1995). This repositioning in G_0–G_1 did not affect the regulation of the transcript levels of the G_1 cyclins (C, D1, D2, D3, and E).

DNA topoisomerases are a unique class of enzymes regulating DNA topology and therefore have been postulated to participate in many cellular processes including DNA repair (Stevnsner and Bohr, 1993). Topoisomerase IIα (p170) predominates in proliferating cells and its protein levels vary during the different phases of the cell cycle, being maximum in G_2–M and undetectable in G_0. Topoisomerase IIβ (p180) is the predominant form in quiescent cells and its protein levels do not vary during the cell cycle (Woessner et al., 1991). The observed decreased levels of the cell cycle-regulated *topoisomerase IIα* transcript may also be a consequence of the effect of radiation in repositioning the cells in G_0–G_1 as the *topoisomerase IIβ* transcript, which is not cell cycle regulated, was not affected.

Increased levels of *gadd45* (growth arrest and DNA damage) mRNA after ionizing radiation treatments (Papathanasiou et al., 1991) occur via a p53-dependent DNA damage response pathway (Kastan et al., 1992; Zhan et al., 1993). In this pathway, the increase in the p53 protein levels is associated with the occurrence of DNA strand breaks (Nelson and Kastan, 1994). *fac1* is a defective gene in Fanconi's anemia complementation group C (Strathdee et al., 1992). These cells are hypersensitive to DNA damage and it was postulated that *fac1* has a role in the protection against, or repair of, DNA damage (Strathdee and Buchwald, 1992; Gordonsmith and Rutherford, 1991). The transcript level changes that were detected occurred after a low-dose-rate irradiation of plateau-phase AG1522, which rendered these cells more radioresistant (adaptive response) to a second challenging dose (Azzam et al., 1992). Whether these changes are an active part of the adaptive response or are an unrelated consequence of the radiation damage remains to be established.

3. ADAPTATION IN C3H 10T½ CELLS

Mouse embryo C3H 10T½ cells have been used by many researchers to study various aspects of neoplastic transformation by ionizing radiation. Similarly, these cells have been used to study the influence of adaptation to radiation on transformation. Azzam et al. (1994a) examined the influence of chronic adapting doses of ^{60}Co γ radiation on MN frequency and on transformation resulting from a subsequent large acute x-ray dose. To reduce the confounding influence of cell cycle effects, confluence arrested cells ($\sim 90\%$ G_0/G_1) were used, as was done in the human fibroblasts. Table I shows representative results. Preexposure of cells

TABLE I
Effect of Chronic Adapting Doses
on Radiation-Induced Micronucleus
Formation and Neoplastic Transformation

Treatment[a]	Percent BNC with MN		Transformation frequency $\times 10^{-3}$ per viable cell
	Observed	Expected	
Control	11.5		0.29
X (4 Gy)	85.3		3.7
γ_1 (0.1 Gy)	16.2		1.5
$\gamma_1 \to X^b$	81.5	90.0	1.7
γ_2 (0.65 Gy)	19.1		0.86
$\gamma_2 \to X$	81.9	92.9	2.5
γ_3 (1.5 Gy)	24.0		0.37
$\gamma_3 \to X$	84.9	97.8	1.7

[a]Adapting doses (γ_1, γ_2, γ_3) were delivered at 0.0024 Gy/min and x-ray test doses were delivered at 2 Gy/min to confluence-arrested cells at 37°C.
[b]$\to$ indicates incubation at 37°C for 3.5 hr.

to chronic doses between 0.1 and 1.5 Gy followed by 3.5 hr of incubation resulted in an adaptive response that protected the cells against both MN formation and neoplastic transformation when these cells subsequently received a second large acute dose. This reduction in MN frequency in the C3H 10T½ cells is the same response that was observed in human fibroblasts. As discussed below, that response involved an increased rate of repair of DNA double strand breaks that result in MN at cell division. The reduction in transformation also seen in the C3H 10T½ cells now indicates further that this increased double strand break repair capacity induced by the pre-exposure must be error free.

It is clear from the data that the magnitude of the protection against transformation (about twofold) was not dose dependent for chronic adapting doses in the range of 0.1 to 1.5 Gy. This indicates, therefore, that adaptation to chronic exposure is already at a maximum after 0.1 Gy and that the change with dose must occur at lower doses. It is important to note, however, that although the magnitude of protection does not increase with dose above 0.1 Gy, neither does it decrease, indicating a stable level of protection up to a large (1.5 Gy) chronic dose. These results therefore raise questions about the validity of current assumptions used for radiation protection, specifically that the cumulative cancer risk from two sequential exposures can never be less than from one alone. Those assumptions do not appear to be consistent with the observed response of these cells.

The transformation experiments described above, and all other data to date describing the adaptive response to radiation, address specifically the question of changes in risk from a subsequent exposure received at times relatively soon after the first adapting exposure. They do not, however, address the consequences for risk resulting from the first exposure alone, i.e., the radiation protection assumption that any single dose results in an increase in risk. However, recent studies do bear on this problem (Azzam et al., 1996). Again, the C3H 10T½ cell line was used, but the cells were grown under conditions that resulted in a high spontaneous transformation frequency. These confluence-arrested cells were exposed to various low doses, at a low dose rate, and were examined for the effect on the transformation frequency. Table II summarizes these results.

The data indicate that doses as low as 0.001 Gy induced a response that reduced the risk of neoplastic transformation in these cells to a rate below that of the original spontaneous level. This protective effect occurred only when the cells were incubated for 24 hr after the exposure, to allow gene expression. The induc-

TABLE II
The Influence of Low Doses on
Spontaneous Transformation Frequency

Treatment[a]	Flasks with foci/ Total flasks	p
Control	34/85	—
0.001 Gy	15/30	0.34
0.001 Gy →[b]	4/27	5.5×10^{-2}
0.01 Gy	16/44	0.47
0.01 Gy →	5/42	9.5×10^{-3}
0.1 Gy	11/42	0.19
0.1 Gy →	6/41	2.3×10^{-2}

[a]All doses delivered at 0.0024 Gy/min to confluence-arrested cells. Radiation exposure produced no significant changes in survival.
[b]→ indicates 24-hr incubation at 37°C before replating.

ing doses spanned two orders of magnitude, yet all produced a similar 3- to 4-fold reduction in the risk of transformation. At these very low doses, it is worthwhile to consider radiation events at the individual cell level. At high doses, all cells receive multiple "hits" from ionizing radiation; that is, many photons or particles pass through each cell, depositing energy along their path and creating a track of ionized molecules and free radicals, mainly derived from water molecules. These in turn react with and create the damage in cellular molecules including DNA. At low doses typical of occupational or background exposure, however, each cell receives many less hits. At a dose of about 0.001 Gy, any cells hit would only be hit about once and therefore experience only one track of damage. Consequently, this is the lowest possible dose that a cell can receive. Our results (Table II) indicate that this dose, which produces a single track through a cell, is sufficient to produce the maximum protective effect, and that at doses up to 100-fold higher, where each cell experiences multiple tracks, this maximum effect is maintained.

As exposure of cells to a low chronic dose protected against both spontaneous and radiation-induced transformation, and the latter appears to be associated with an enhanced error-free type of double strand break repair, it is possible that spontaneous transformation may share a common critical process with radiation-induced transformation, and that process may involve DNA double-strand breaks and their repair.

4. INTERCELLULAR SIGNALS

One of the assumptions associated with the current model of cancer risk is that risk arises only in those cells actually experiencing a radiation track and those cells do not influence the risk of any other cell.

Studies have been done to test this assumption and examine the possibility that a radiation exposure induces cells to communicate with other unexposed cells in such a way as to alter the unexposed cells. Exposure of human lymphocytes to doses at or below 0.001 Gy (one track per cell) enhanced the expression of IL-2α receptors on the lymphocyte surface, when the cells were subsequently stimulated to divide. These irradiated cells also released a factor into the culture medium that stimulated the same response in unirradiated cells (Xu *et al.*, 1996).

These data show that cells exposed to typical occupational doses can respond to that dose by altering the level of surface cytokine receptors and to also communicate with other unexposed cells, causing a similar response in them. Although the significance for risk of these events is as yet unclear, it should be noted that IL-2 is a cytokine typically involved in cell proliferation responses, a process central to tumorigenesis. What is clear, however, is that cell-to-cell communication can be altered by very low doses of radiation.

5. ADAPTATION AND APOPTOSIS

When a normal cell is exposed to radiation, there are three possible outcomes that impact directly on cancer risk in that cell. Repair may take place, returning the cell to its original state and no long-term consequence existing; the repair may be incomplete or inaccurate such that the cell suffers a mutation, which may or may not be a cancer-initiating lesion; the cell may die, which eliminates any cancer risk. Adaptation may influence the first two possibilities as described above, and also have an effect on necrotic cell death in fibroblasts. However, radiation-induced cell death in some cell types, including bone marrow cells which are at particular cancer risk from radiation, is by the alternate process of apoptosis. Human lymphocytes were chosen as representative lymphoid cells to assess the influence of adaptation on radiation-induced apoptosis. A small radiation exposure (0.1 Gy) given to primary unstimulated human lymphocytes 6 hr prior to a large 3-Gy acute test exposure significantly increased, rather than decreased, the probability that the cells would subsequently die by apoptosis (Cregan et al., 1994). This result appears contradictory to previous data indicating that adaptive processes in lymphocytes (Wiencke et al., 1986) as well as fibroblasts (Azzam et al., 1992) protect the cell from the consequences (including chromosomal aberrations and deletions) of a further exposure. However, if the adaptive response exists to ultimately protect the whole organism, as seen for example in the reduction of transformation frequency in the C3H 10T½ cells, then the increased sensitivity to radiation-induced apoptosis makes sense at the level of the whole organism. Presumably any cells experiencing DNA repair problems would be more likely to die, reducing the probability of survival with a tumor-initiating mutation. It is not clear whether the reduction in chromosomal aberrations seen by Wiencke *et al.* (1986) is in part or in whole related to apoptotic loss of cells with misrepaired DNA. This increased sensitivity to radiation-induced apoptosis is not unique to radiation-induced adaptation. Lymphocytes that experienced a very mild hyperthermia stress (40°C, 30 min) 12 hr prior to the large acute radiation test exposure were also more likely to die by apoptosis than unheated controls (Boreham *et al.*, 1996a). These results are consistent with the other hyperthermia experiments described above, indicating that a heat stress is capable of inducing the adaptive process. They also indicate that a very mild heat stress, well within the common temperature range of human fevers and of even shorter duration, can markedly alter cellular risk from radiation exposure. These observations are therefore relevant and important for both occupationally exposed individuals who may be returning to work after a short fever, and patients receiving cancer therapy who may be ill for a variety of reasons.

6. MECHANISMS

Measurements have been performed on the relative rates of repair of DNA double strand breaks that lead to MN in adapted and unadapted human skin fibroblasts, by holding the cells in confluence at 37°C for 2 hr after exposure to the acute test dose (Table III) (Azzam *et al.*, 1994b). During this holding period, both adapted and unadapted cells repaired breaks and the ultimate MN frequency decreased. However, adapted cells were able to repair a greater fraction of their double strand breaks in this period than unadapted cells (about one-third greater), indicating that the rate of double strand break repair was increased in radiation-adapted cells. Extension of the incubation period to longer times than 2 hr for both the adapted and nonadapted cells ultimately resulted in the same maximum reduction in MN frequency. These results indicate that although the rate of double-strand break repair was increased by adaptation to radiation, the ultimate level of repair, given sufficient time, was not changed. This finding suggests that DNA double-strand breaks that are unrepairable in adapted cells are also unrepairable in nonadapted cells, further suggesting that adaptation results from induction of a DNA double-strand break repair system already functioning at a constitutive level.

TABLE III
Rate of Decrease in MN Frequency
of Adapted and Unadapted Normal
Human Skin Fibroblasts

Treatment[a]	MN frequency $\pm$ SE	Frequency decrease (%)
3 Gy	0.97 ± 0.04	—
3 Gy $\overset{2h}{\to}$[b]	0.68 ± 0.03	30 ($p < 0.001$)
0.5 Gy $\overset{0h}{\to}$ 3 Gy	0.83 ± 0.03	—
0.5 Gy $\overset{0h}{\to}$ 3 Gy $\overset{2h}{\to}$	0.49 ± 0.02	41 ($p < 0.001$)

[a]The 0.5-Gy adapting dose was delivered chronically at 0.0005 Gy/min to cells at 37°C and the 3-Gy test dose acutely at 2 Gy/min.
[b]$\to$ indicates cells were held at 37°C.

Changes in radiation-induced division delay in adapted cells were also estimated, by determining the number of cells that reach the binucleate state within a specified time, when incubated in the presence of cytochalasin B (to allow nuclear, but prevent cellular division) after a radiation exposure. Table IV shows results obtained using rodent C3H 10T½ cells (Mitchel, 1995), and similar results have been obtained in human skin fibroblasts (unpublished observations). The results for the C3H 10T½ cells show that although a rather large chronic γ-ray exposure (1.5 Gy at 0.0025 Gy/min) does not by itself alter division rates compared with unexposed cells, this adapting dose significantly increases division delay in cells subsequently exposed to a 4-Gy acute x-ray test

TABLE IV
Increased Division Delay
in Radiation-Adapted Confluent
C3H 10T½ Cells

Treatment	Binucleate cells (%)
Control	46.2 ± 2.5
X (4 Gy acute)	21.3 ± 1.5
γ (1.5 Gy chronic)	45.3 ± 1.8
γ + X	11.0 ± 1.1

dose, as compared with the division delay produced by the 4-Gy acute test dose alone. In this experiment cells were incubated for 48 hr in cytochalasin B, and survival after the combined doses was not different from the acute dose alone.

In general, results described in the sections above have indicated that radiation-exposed cells preadapted to radiation show increased rates of DNA double-strand break repair and increased division delay, leading to a reduced level of chromosomal breaks and a reduced probability of transformation. These observations are consistent with induced error-free recombinational repair, and error-free recombinational repair is considered to occur between homologous chromosomes. Recombination between homologous chromosomes further implies that these chromosomes must pair or align in some way in response to radiation, to allow that recombinational repair process to occur. To test this hypothesis, FISH technology has been used to fluorescently stain homologous chromosome pairs (chromosomes 7 and 21 in interphase nuclei of attached confluent human fibroblasts and endothelial cells). Confocal microscopy was then used to assess the relative distance between the two chromosome 7 or 21 domains in each nucleus. The data indicate that 2 hr after an acute 4-Gy exposure, about 70% of nuclei contained chromosome 21 domains close together (<4 µm apart) as compared with unexposed cells, where only about 34% were close together. Similar results were obtained for chromosome 7 domains and when a lower dose of 0.5 Gy was used (Dolling *et al.*, 1997). These results support the hypothesis of a radiation-induced pairing of homologous chromosomes, a condition likely to facilitate error-free recombinational repair. The possibility that radiation-adapted cells may have increased chromosomal pairing is being pursued.

Recently, Lehnert and Chow (1996) reported that, in a cell-free system, nuclear extracts of radiation-adapted human cells were better able to catalyze homologous double-strand recombination between deletion plasmids, as compared with unadapted cells, further supporting the idea that adaptation to radiation enhances cellular capacity for homologous recombination.

7. COMPARISON WITH A LOWER EUKARYOTE

The ability to adapt to ionizing radiation and increase cellular radioresistance is well known in yeast, and has now been demonstrated in mammalian cells. This, then, raises the question whether the mammalian system that is induced by radiation is basically the same evolutionarily conserved system that exists in yeast (Mitchel and Morrison, 1982, 1984, 1987). Table V compares the known characteristics of each, and shows that there is a close correlation between the two systems. Adaptation to radiation in yeast is known to be based on the induction of homologous recombinational repair of DNA (Mitchel and Morrison, 1982). The characteristics of the mammalian system appear to be consistent with the same process.

8. CONCLUSIONS AND IMPLICATIONS

It now seems clear that human and other mammalian cells capable of normal division are able to adapt in ways that temporarily alter their response to and reduce the consequences of a radiation exposure. These cells appear to respond on virtually an all-or-nothing basis, as a dose equivalent to about one track per cell (0.001 Gy) appears to produce a maximum response. On the other hand, increasing the dose up to very large total doses (4 Gy) delivered at a low dose rate maintains the response but appears to

TABLE V
Adaptive Response to Ionizing Radiation

	Mammals	Yeast
Radioresistance induced by		
radiation	Yes	Yes
Other DNA-damaging agents	Yes	Yes
Heat	Yes	Yes
Survival increased	Yes	Yes
Response proportional to dose	No	Yes
Maximum response reached	Yes	Yes
Adaptation requires metabolism	Yes	Yes
Double-strand break repair induced	Yes	Yes
Induced repair is error free	Yes?	Yes
Induced repair is homologous recombinational repair	?	Yes
Induced repair recognizes nonradiation DNA damage	Yes	Yes
Single-strand break repair induced	No	No
Division delay induced	Yes	?
DNA "damage" is a signal	?	Yes

have no further influence on this maximum effect. Dose rate, however, does influence the response. Cells are induced to a higher level of resistance after chronic rather than acute doses. Considering the all-or-nothing response at low dose rates, this result suggests that high-dose-rate exposures must induce competing processes that in some cells prevent adaptation.

The processes and responses to radiation observed in these mammalian cells appear to bear directly on the current assumptions made about the risk of a radiation exposure, assumptions based on the linear no-threshold model (ICRP, 1991). That model predicts that incremental low doses of radiation produce proportional and incremental increases in cancer risk, and that every dose, no matter how low, increases cancer risk. The predictions of that model form the basis of accepted principles of radiation protection, and the large amounts of money and concern used to reduce population and individual exposure to ionizing radiation are justified on that model's prediction of risk. The responses of human and rodent cells to low and chronic radiation exposures described here, however, raise questions about the validity of the linear model under the circumstances described. The data indicating that the risk of chromosomal damage and transformation can decrease with incremental doses do not fit the model. The data indicating that a single low dose (as low as one track per cell) reduces rather than increases the risk of spontaneous transformation raise questions about the assumption that all doses, no matter how small, increase risk. Finally, the observation that another stress, mild heat, can induce the same adaptation to radiation raises questions about a constant value for risk in relation to dose.

There are, however, still many uncertainties that must be addressed before the significance of these adaptive processes for human cancer risk can be determined. As noted earlier, several investigators have reported an inability to detect adaptation to radiation in lymphocytes of some individuals or in cells from some mouse strains. It may be that adaptation to radiation is a genetic variable and that this response, or its absence, contributes to the heterogeneity seen in the radiation response of the human population. Alternatively, our ability to detect or stimulate the response may depend on other poorly understood variables. The observations presented in this chapter indicate that dose and dose rate are important parameters for induction, and there may be individual or strain differences in sensitivity to these variables. The data also indicate that other stresses like heat will induce radiation resistance. Inadvertent mild, short heat stress (and possibly other stresses) could induce the full extent of radioresistance prior to exposure to the experimental adapting dose of radiation, precluding any further change in response to that dose. As a consequence the cells could appear unresponsive. Another major uncertainty is the ability of cells from different organs to adapt to radiation. Only a few cell types have been tested. Little is known about the relative impact of adaptation on cancer risk in cells of different tissue origin. Little is also known about cell cycle influences on the ability of cells to adapt, or on the duration of the adapted state. The work presented here on lymphocytes, fibroblasts, and rodent embryo cells was deliberately done on confluence-arrested cells to avoid potentially confounding cell cycle influences. Under those circumstances, adaptation lasted for at least 24 hr. Many cells in adult humans are typically in a nondivisional state, although some are clearly otherwise. Bone marrow cells for example are normally dividing, and also represent a cell type at high risk for radiation-induced cancer. The existence, duration, and significance for cancer risk of adaptation to radiation in these cells may be particularly important, and at this time is certainly unclear.

Because the biological responses of cells to low radiation doses described here are neither considered by, nor appear to be consistent with the current model of risk prediction, it appears important that research efforts be made to assess the impact of the biology of low dose responses on that model.

ACKNOWLEDGMENT. This work was supported by the CANDU Owners Group.

REFERENCES

Azzam, E. I., de Toledo, S. M., Raaphorst, G. P. and Mitchel, R. E. J., 1992, Radiation-induced radioresistance in a normal human skin fibroblast cell line, in: *Low Dose Irradiation and Biological Defense Mechanisms* (T. Sugahara, L. A. Sagan, T. Aoyama, eds.), Elsevier, Amsterdam, pp. 291–294.

Azzam, E. I., Raaphorst, G. P., and Mitchel, R. E. J., 1994a, Radiation-induced adaptive response for protection against micronucleus formation and neoplastic transformation in C3H 10T½ mouse embryo cells, *Radiat. Res.* **138:** S28–S31.

Azzam, E. I., de Toledo, S. M., Raaphorst, G. P., and Mitchel, R. E. J., 1994b, Réponse adaptative au rayonnement ionisant des fibroblastes de peau humaine. Augmentation de la vitesse de réparation de l'ADN et variations de l'expression des génes, *J. Chim. Phys. Phys. Chim. Biol.* **91:**931–936.

Azzam, E. I., de Toledo, S. M., Raaphorst, G. P., and Mitchel, R. E. J., 1996, Low-dose ionizing radiation decreases the frequency of neoplastic transformation to a level below the spontaneous rate in C3H 10T½ cells, *Radiat. Res.* **146:** 369–373.

Boreham, D. R., and Mitchel, R. E. J., 1991, DNA lesions that signal the induction of radioresistance and DNA repair in yeast, *Radiat. Res.* **128:**19–28.

Boreham, D. R., and Mitchel, R. E. J., 1993, DNA repair in *Chlamydomonas reinhardtii* induced by heat shock and gamma radiation, *Radiat. Res.* **135:** 365–371.

Boreham, D. R., and Mitchel, R. E. J., 1994, Regulation of heat and radiation stress responses in yeast by hsp-104, *Radiat. Res.* **137:**190–195.

Boreham, D. R., Trivedi, A., Weinberger, P. and Mitchel, R. E. J.,1990, The involvement of topoisomerases and DNA polymerase I in the mechanism of induced thermal and radiation resistance in yeast, *Radiat. Res.* **123:**203–212.

Boreham, D. R., Trivedi, A., and Mitchel, R. E. J., 1991, Radiation and stress response in *Saccharomyces cerevisiae*, in: *Molecular Biology of Yeast in Relation to Biotechnology* (R. Prasad, ed.), Omega Scientific, New Delhi, India, pp. 295–314.

Boreham, D. R., Walker, J.-A., Maves, S. R., Greiner, K., Mitchel, R. E. J., and Lucas, J. N., 1994, Mild hyperthermia-induced modification of radiation induced chromosomal aberrations. A comparison of micronuclei and translocations, *Proceedings of the 42nd Annual Meeting of the Radiation Research Society*, Abstract P03-51, p. 119.

Boreham, D. R., Maves, S. R., Miller, S., Morrison, D. P., and Mitchel, R. E. J., 1996, Heat induced thermal tolerance and radiation resistance to apoptosis in human lymphocytes, *Proceedings of the 44th Annual Meeting of the Radiation Research Society*, Abstract P19-328, p. 156.

Calkins, J., 1967, An unusual form of response in x-irradiated protozoa and a hypothesis as to its origin, *Int. J. Radiat. Biol.* **4:**297–301.

Cregan, S. P., Boreham, D. R., Walker, J.-A., Brown, D. L., and Mitchel, R. E. J., 1994, Modification of radiation-induced apoptosis in radiation or hyperthermia-adapted human lymphocytes, *Biochem. Cell Biol.* **72:**475–482.

de Toledo, S. M., Azzam, E. I., Gasmann, M. K., and Mitchel, R. E. J., 1995, Use of semi-quantitative reverse transcription-polymerase chain reaction to study gene expression in normal skin fibroblasts following low dose-rate irradiation, *Int. J. Radiat. Biol.* **67:**135–142.

Dolling, J.-A., Boreham, D. R., Brown, D. L., Raaphorst, G. P., and Mitchel, R. E. J., 1997, Rearrangement of human cell homologous chromosome domains in response to ionizing radiation, *Int. J. Radiat. Biol.* **72:**303–311.

Dulic, V., Kaufmann, W. K., Wilson, S. J., Tisty, T. D., Lees, E., Wade Harper, J., Elledge, S. J., and Reed, S. I., 1994, P53-dependent inhibition of cyclin-dependent kinase activities in human fibroblasts during radiation-induced G_1 arrest, *Cell* **76:**1013–1023.

Gordon-Smith, E. C., and Rutherford, T. R., 1991, Fanconi anemia: Constitutional aplastic-anemia, *Semin. Hematol.* **28:**104–112.

ICRP, 1991, 1990 Recommendations of the International Commission on Radiological Protection, Publication 60, *Annals of the ICRP*, 21, Pergamon Press, London.

Kastan, M. B., Zhan, Q., El-Deiry, W. S., Carrier, F., Jacks, T., Walsh, W. V., Plunkett, B. S., Volgestein, B., and Fornace A. J., Jr., 1992, A mammalian cell cycle checkpoint pathway utilizing p53 and GADD45 is defective in ataxia-telangiectasia, *Cell* **71:**587–597.

Koval, T. M., 1986, Inducible repair of ionizing radiation damage in higher eukaryotic cells, *Mutat. Res.* **173:**291–293.

Koval, T. M., 1988, Enhanced recovery from ionizing radiation damage in a lepidopteran insect cell line, *Radiat. Res.* **115:**413–420.

Lambin, P., Marples, B., Malaise, E. P., Fertil, B., and Joiner, M. C., 1993, Hypersensitivity of a human tumour cell line to very low radiation doses, *Int. J. Radiat. Biol.* **63:**639–650.

Lambin, P., Fertil, B., Malaise, E. P., and Joiner, M. C., 1994, Multiphasic survival curves for cells of human tumor cell lines: Induced repair or hypersensitive subpopulation? *Radiat. Res.* **138:**S32–S36.

Lehnert, S., and Chow, T. Y. K., 1996, Low doses of ionizing radiation induce nuclear activity catalyzing double strand homologous DNA recombination, *Proceedings of the 44th Annual Meeting of the Radiation Research Society,* Abstract P14-262, p. 140.

Marples, B., and Joiner, M. C., 1993, The response of Chinese hamster V79 cells to low radiation doses: evidence of enhanced sensitivity of the whole population, *Radiat. Res.* **133:**41–51.

Marples, B., Joiner, M. C., and Skov, K., 1994, The effect of oxygen on low-dose hypersensitivity and increased radioresistance in Chinese hamster V79-379A cells, *Radiat. Res.* **138:**S17–S20.

Mitchel, R. E. J., 1995, Mechanisms for the adaptive response in irradiated mammalian cells, *Radiat. Res.* **141:**117–118.

Mitchel, R. E. J., and Morrison, D. P., 1982, Heat shock induction of ionizing radiation resistance in *Saccharomyces cerevisiae.* Transient changes in growth cycle distribution and recombinational ability, *Radiat. Res.* **92:**182–187.

Mitchel, R. E. J., and Morrison, D. P., 1984, An oxygen effect for gamma-radiation induction of radiation resistance in yeast, *Radiat. Res.* **100:**205–210.

Mitchel, R. E. J., and Morrison, D. P., 1987, Inducible DNA-repair systems in yeast: Competition for lesions, *Mutat. Res.* **183:**149–159.

Mushel, R. J., Zhang, H. B., Iliakis, G., and McKenna, W. G., 1991, Cyclin B expression in HeLa cells during the G2 block induced by ionizing radiation, *Cancer Res.* **51:**5113–5117.

Nelson, W. G., and Kastan, M. B., 1994, DNA strand breaks: The DNA template alterations that trigger p53-dependent DNA damage response pathways, *Mol. Cell. Biol.* **14:**1815–1823.

Olivieri, G., and Bosi, A., 1990, Possible causes of variability of the adaptive response in human lymphocytes, in: *Chromosomal Aberrations: Basic and Applied Aspects* (G. Obe and A.T. Natarajan, eds.), Springer-Verlag, Berlin, pp. 130–139.

Olivieri, G., Bodycote, J., and Wolff, S., 1984, Adaptive response of human lymphocytes to low concentration of radioactive thymidine, *Science* **23:**594–597.

Pagano, M., Pepperkok, P., Verde, F., Ansorge, W., and Draetta, G., 1992, Cyclin A is required at two points in the human cell cycle, *EMBO J.* **11**:961–971.

Papathanasiou, M. A., Kerr, N. C. K., Robbins, J. H., McBride, O. W., Alamo, I., Jr., Barret, S. F., Hickson, I. D., and Fornace, A. J., Jr., 1991, Induction by ionizing radiation of the GADD45 gene in cultured human cells: Lack of mediation by protein kinase C, *Mol. Cell. Biol.* **11**:1009–1016.

Resnitzky, D., Gossen, M., Bujard, H., and Reed, S. I., 1994, Acceleration of the G_1/S transition by expression of cyclins D1 and E using an inducible system, *Mol. Cell. Biol.* **14**:1669–1679.

Shadley, J. D., and Wiencke, J. K., 1989, Induction of the adaptive response by x-rays is dependent on radiation intensity, *Int. J. Radiat. Biol.* **56**:107–118.

Sherr, C. J., 1993, Mammalian G_1 cyclins, *Cell* **73**:1059–1065.

Spadinger, I., Marples, B., Mathews, J., and Skov, K., 1994, Can colony size be used to detect low-dose effects? *Radiat. Res.* **138**:S21–S24.

Stevnsner, T., and Bohr, V., 1993, Studies on the role of topoisomerases in general, gene- and strand-specific DNA repair, *Carcinogenesis* **14**:1841–1850.

Strathdee, C. A., and Buchwald, M., 1992, Molecular and cellular biology of Fanconi anemia, *Am. J. Pediatr. Hematol. Oncol.* **14**:177–185.

Strathdee, C. A., Gavish, H., Shannon, W. R., and Buchwald, M., 1992, Cloning of cDNAs for Fanconi's anaemia by functional complementation, *Nature* **356**: 763–767.

Walker, G. C., 1984, Mutagenesis and inducible responses to deoxyribonucleic acid damage in *Escherichia coli, Microbiol. Rev.* **48**:60–93.

Walker, G. C., 1985, Inducible DNA repair systems, *Annu. Rev. Biochem.* **54**: 425–457.

Wiencke, J. K., Afzal, V., Olivieri, G., and Wolff, S., 1986, Evidence that the [^{3}H]-thymidine-induced adaptive response of human lymphocytes to subsequent doses of X-rays involves the induction of a chromosomal repair mechanism, *Mutagenesis* **1**:375–380.

Wiencke, J. K., Shadley, J. D., Kelsey, K. T., Kronenberg, A., and Little, J. B., 1987, Failure of high intensity X-ray treatments or densely ionizing fast neutrons to induce the adaptive response in human lymphocytes, in: *Proceedings of the 8th International Congress of Radiation Research* (E. M. Fielden, J. F. Fowler, J. H. Hendry, and D. Scott, eds.), Taylor & Francis, London, Vol. 1, p. 212.

Wilson, A., 1989, Meeting report: The effects of small doses of radiation, *Int. J. Radiat. Biol.* **56**:203–206.

Woessner, R. D., Mattern, M. R., Mirabelli, C. K., Johnson, R. K., and Drake, F. H., 1991, Proliferation- and cell cycle-dependent differences in expression of the 170 kilodalton and 180 kilodalton forms of topoisomerase II in NIH-3T3 cells, *Cell Growth Differ.* **2**:209–214.

Wojcik, A., and Tuschl, H., 1990, Indications of an adaptive response in C57BL mice pre-exposed in vivo to low doses of ionizing radiation, *Mutat. Res.* **243**: 67–73.

Wojcik, A., Bonk, K., Müller, W.-U., Streffer, C., Weissenborn, U., and Obe, G., 1992a, Absence of adaptive response to low doses of X-rays in preimplantation embryos and spleen lymphocytes of an inbred mouse strain as compared to human peripheral lymphocytes: A cytogenetic study, *Int. J. Radiat. Biol.* **62**: 177–186.

Wojcik, A., Bonk, K., Müller, W.-U., and Streffer, C., 1992b, Indications of strain specificity for the induction of adaptive response to ionizing radiation in mice, in: *Low Dose Irradiation and Biological Defense Mechanisms* (T. Sugahara, L. A. Sagan, and T. Aoyama, eds.), Elsevier, Amsterdam, pp. 311–314.

Wolff, S., 1992a, Failla Memorial Lecture: Is radiation all bad? The search for adaptation, *Radiat. Res.* **131:**117–123.

Wolff, S., 1992b, Low dose exposure and the induction of adaptation, in: *Low Dose Irradiation and Biological Defense Mechanisms* (T. Sugahara, L. A. Sagan, and T. Aoyama, eds.), Elsevier, Amsterdam, pp. 21–28.

Wolff, S., Afzal, V., Wiencke, J. K., Olivieri, G. and Micheali, A., 1988, Human lymphocytes exposed to low doses of ionizing radiations become refractory to high doses of radiation as well as chemical mutagens that induce double-strand breaks in DNA, *Int. J. Radiat. Biol.* **53:**39–48.

Wolff, S., Wiencke, J. K., Afzal, V., Youngblom, J., and Cortés, F., 1989, The adaptive response of human lymphocytes to very low doses of ionizing radiation: A case of induced chromosomal repair with the induction of specific proteins, in: *Low Dose Radiation: Biological Bases of Risk Assessment* (K. F. Baverstock and J. W. Stather, eds.), Taylor & Francis, London, pp. 446–454.

Wouters, B. G., and Skarsgard, L. D., 1994, The response of a human tumor cell line to low radiation doses: Evidence for enhanced sensitivity, *Radiat. Res.* **138:**S76–S80.

Xiong, Y., and Beach, D., 1991, Population explosion in the cyclin family, *Curr. Biol.* **1:**362–364.

Xu, Y., Greenstock, C. L., Trivedi, A., and Mitchel, R. E. J., 1996, Occupational levels of radiation exposure induce surface expression of interleukin-2 receptors in stimulated human peripheral blood lymphocytes, *Radiat. Environ. Biophys.* **35:**89–93.

Zhan, Q., Carrier, F., and Fornace, A. J., Jr., 1993, Induction of cellular p53 activity by DNA-damaging agents and growth arrest, *Mol. Cell. Biol.* **13:**4242–4250.

Index

When page numbers are followed by *f* or *t*, readers can find the subject cited in the figure or table, respectively, on that page.